HOW TO FIND OUT
IN MATHEMATICS

A guide to sources of mathematical
information arranged according to
the Dewey Decimal Classification

by

JOHN E. PEMBERTON
B.A., F.L.A.

PERGAMON PRESS

OXFORD · LONDON · PARIS · FRANKFURT

THE MACMILLAN COMPANY
NEW YORK

PERGAMON PRESS LTD.
Headington Hill Hall, Oxford
4 & 5 Fitzroy Square, London W.1

THE MACMILLAN COMPANY
60 Fifth Avenue, New York 11, New York

COLLIER-MACMILLAN CANADA, LTD.
132 Water Street South, Galt, Ontario, Canada

GAUTHIER-VILLARS ED.
55 Quai des Grands-Augustins, Paris 6

PERGAMON PRESS G.m.b.H.
Kaiserstrasse 75, Frankfurt am Main

Library of Congress Card No. 63–22292

Set in 10 on 12 pt Times and printed in Great Britain by
DITCHLING PRESS LTD · HASSOCKS · SUSSEX

Contents

TO MY WIFE

Preface

THE purpose of this book is to give to those studying or practising
mathematics an insight into the vast amount of information which
is available, and to act as a guide to its exploitation. A balance has
been sought between descriptions of actual sources and the
principles of library and information work.

It is strongly recommended that the whole text is first read
through from beginning to end before detailed attention is given
to any individual chapter or section. It is extremely important to
see the total picture in perspective before examining the minute
detail of its constituent parts. Whilst the text has been designed to
facilitate continuous reading, the book has considerable reference
value, which is enhanced by a detailed index.

Used in this way the work is suitable for students and practitioners at all levels: from the higher grades of secondary school,
through college and university, and up to such highly specialized
fields as operational research and actuarial science.

The chapters are arranged according to the Mathematics class
of the Dewey Decimal Classification which, apart from providing
a convenient and progressive grouping of material, also familiarizes
the reader with the classification scheme most likely to be en-
countered in libraries and works of reference.

Sources of Russian mathematical information are treated
separately in an appendix because of their unique appeal, whilst
contributions from other countries are incorporated in the body
of the book. Both actuarial science and the role of government

departments are similarly dealt with as appendices.

I wish to record my sincere appreciation of the willing co-operation I have received from the following bodies: American Statistical Association; Association of Teachers in Colleges and Departments of Education; Industrial Mathematics Society; Institute of Actuaries; London Mathematical Society; Mathematical Association; Mathematical Association of America; Operational Research Society; Operations Research Society of America; Royal Statistical Society; Society for Industrial and Applied Mathematics; U.S. Bureau of Labor Statistics.

I am also indebted to Dr George Chandler, M.A., F.L.A., F.R.Hist.S., City Librarian of Liverpool, who read the manuscript and proofs and made some useful suggestions.

Liverpool   JOHN E. PEMBERTON

CHAPTER 1

Careers for Mathematicians

'With the post-war eruption in mathematics the problem of supply [of mathematicians] has become a serious one, and in some countries has taken on the character of a political as well as a social crisis."

E. R. Duncan, in the *Year Book of Education*, Evans, London, 1961.

General Survey

Before the question of careers can be discussed it is necessary to define the word *mathematician*. Mathematics is an essential training for many scientific and technical as well as commercial and business occupations, but to attempt to cover all the fields in which some mathematical knowledge is required is beyond the scope of the present work. The mathematician is here considered as a person who has specialized in mathematics at degree level.

Approximately one-half of each year's product of mathematicians find employment in teaching of one form or another. The other half are spread through a wide range of situations. Qualified mathematicians are indeed to be found in almost every sector of industry, though, as might be expected, many are grouped in electronics, aeronautics, and space exploration.

Two developments which have in recent years opened up bright new horizons for mathematicians, and which will continue to offer good prospects for many years to come, are high-speed computers and operational research.

The now established concept of automation has affected many industries and brought in its wake new openings for those with the necessary skills. Industries which are particularly concerned in this development include chemical processing, nuclear engineering,

1

iron and steel and, of course, those which produce the necessary control systems.

The specialized field of actuarial science is facing growing competition for new recruits, but continues to offer good prospects for suitably qualified men and women.

Patterns of mathematical employment are constantly changing, but they are accurately reflected in the reports of the Mathematical Sciences Section of the National Register of Scientific and Technical Personnel. This is maintained by the American Mathematical Society in co-operation with the National Science Foundation. Ten other societies, representing such fields as statistics, electronic computation, actuarial science, and operational research, assist in the work. Data covering the year 1960 are analysed in a report entitled: *Professional Characteristics of Mathematical Scientists*, American Mathematical Society, Providence, R.I., 1962. Among the questions with which the report deals are employment trends among younger mathematicians, patterns of degree attainment, the proportion of qualified persons absorbed by the individual subfields of mathematics, and salaries. Two especially interesting inferences are drawn which are equally valid for many countries outside the United States. The first is the "enormous growth of opportunities for non-academic employment of mathematicians"; and the second is that "mathematicians in industry, business and non-profit organizations are getting higher salaries at earlier ages than those in other types of employment". The discernible shift away from teaching is largely due to a strong attraction from new specialities which do not demand as high a level of academic attainment.

It will be stated on several occasions in this book that the work of the mathematician is often geared to the task of helping to solve the problems of workers in apparently unrelated fields. The contribution which the application of mathematical techniques can make in terms of efficiency and hard cash has created an unprecedented demand for the services of competent mathematicians. They are now being taken on in industry for their knowledge of mathematics itself rather than of the fields to which it is applied. It is significant that in the United States the number of mathematicians employed in private industry increased by nearly 100 per cent between 1954 and 1960. Organizations both large and small have found that their volume of experimentation can be considerably reduced by the elaboration of mathematical formulations. The aircraft industry is one which has particularly benefited in this respect. However, many less obvious areas of research are also directly concerned in this movement.

It is the purpose of this chapter to draw attention to the main areas of employment, giving at the same time a brief guide to sources of relevant information.

Guides to Careers

Guides to careers in mathematics appear under three main headings: general guides to occupations for scientists, particularly graduates; those dealing with the entire range of mathematical opportunities; and very specific guides which restrict themselves to individual mathematical professions. The following examples represent the first and second categories. More detailed information on particular professions is given in succeeding sections.

ANGEL, JUVENAL LONDOÑO, *Careers for Majors in Mathematics.* World Trade Academy Press, New York, 1959.

Bibliography of Current Occupational Literature. National Vocational Guidance Association, Washington, D.C., 1956.

Careers in Mathematics. Institute for Research, Chicago, 1959.

Directory of Opportunities for Qualified Men. Cornmarket Press, London, 1961.

FORRESTER, GERTRUDE, *Occupational Literature: an Annotated Guide.* Wilson, New York, 1958.

Jobs in Mathematics. Science Research Associates, Chicago, 1959.

The Mathematician. H.M.S.O., London, 1961 (Series: Choice of Careers, No. 109).

Professional Opportunities in Mathematics. Mathematical Association of America, Buffalo, 1961. (Prepared by the Association's Committee on Advisement and Personnel. The booklet is an outgrowth of an article which first appeared in the *American Mathematical Monthly*, and is now in its 5th edition.)

SHEDD, A. NEAL *et al. Careers in Engineering, Mathematics, Science and*

Related Fields: a Selected Bibliography. U.S. Office of Education, 1961.
TURNER, NURA D., *A Bibliography for Careers in Mathematics.* In *Science Education*, Oct. 1961, and *Mathematics Teacher*, Nov. 1961.

A pamphlet on careers in mathematics has been prepared by and is available from the Conference Board of the Mathematical Sciences, 1515 Massachusetts Avenue, N.W., Washington 5, D.C., U.S.A. The first section deals with teaching careers and the second with industrial and government careers in mathematics.

A very useful survey was published by the U.S. Bureau of Labor Statistics under the title "Professional mathematicians work in industry and government", *Monthly Labor Review*, Sept. 1961. This is available as Reprint No. 2374 from the Bureau of Labor Statistics, Room 1025, 341 9th Avenue, New York 1, New York, U.S.A.

There are a number of periodicals on careers which may be consulted as they frequently contain articles on mathematical openings. They include the *Journal of College Placement*, *Tomorrow's Scientists*, and *Occupational Outlook Quarterly*. The last named, for example, carried an article by George Hermanson on "Employment in professional mathematical work in industry and government" in its December 1961 issue.

Employment Registers

An employment register is jointly maintained by the American Mathematical Society, the Mathematical Association of America and the Society for Industrial and Applied Mathematics. The Employment Register, whose services are not limited to members of these bodies, contains a list of positions available to mathematicians in industry, government, and academic institutions, and a list of applicants.

In Britain, the national advisory body is the Technical and Scientific Register, Ministry of Labour, Almack House, 26–28 King Street, London, S.W.1.

Most, if not all, universities and colleges offer positive assistance in finding suitable employment to those completing their studies. Often, the placement officer is well known to local employers who notify him of vacancies. Those institutions which have a high reputation for producing good mathematicians have valuable employment contacts on a national, even an international basis. It is a good plan to peruse the advertisements in mathematical and other journals for some months before graduation in order to get the feel of the market. Such reading can give a helpful stimulus to a student preparing for his final examinations.

Teaching and Women

Teaching mathematics will be dealt with in Chapter 6, but a word may be added here on the employment of women. Married women who hold academic qualifications are being increasingly sought to help meet the present-day shortage of teachers. Several countries have launched campaigns to attract married women back into the profession.

Among newly qualified women teaching is a favoured occupation. In the United States in 1957, 1166 women were graduated with an undergraduate major in mathematics. Of these, 86 per cent took jobs, and within this group 42 per cent were engaged as teachers. By way of comparison, another 42 per cent were engaged as mathematicians or statisticians, and 16 per cent in a variety of other positions. The National Register of Scientific and Technical Personnel for 1956–58 showed 1277 women in mathematical occupations. This was over 10 per cent of the total mathematical recordings. About 50 per cent were employed by colleges and universities, 30 per cent by industry, and 20 per cent by government agencies and other bodies. The figure for 1960 showed a leap to 1633 women mathematicians, most of whom did not have higher degrees. It is reasonable to assume that an increasingly high proportion are finding employment as computer programmers.

Part-time abstracting work can often be an attractive proposition. It is quite well paid, particularly when the abstractor is capable of handling material in one or more foreign languages in her field of specialization. The Management Services Group of the National Federation of Science Abstracting and Indexing Services (301 East Capitol Street, Washington 3, D.C.) and Aslib (3 Belgrave Square, London, S.W.1) can give advice on opportunities.

Research

In the field of research, mathematics is often referred to as *the language of science*. This is because it is through mathematical formulations used in the different branches of science that workers in one can understand the problems and progress of others, and can in turn communicate their own. Mathematics is the common medium through which research can progress. The mathematician performs the necessary function of formulating problems, of rationalizing them in a manner which will most readily render them capable of solution. In this he frequently uses the technique of the mathematical model.

Recently, a leading company in the propulsion field advertised for a mathematician "to develop programmes that transform design intent into the definitions required for numerically controlled tools. Will mathematically define rocket engine compounds. A degree in mathematics is required." This is typical of the current industrial demand.

As industrial developments shift their emphasis and change their course, the one constant need will be for men soundly versed in the theoretical disciplines—especially mathematics. In 1961, UNESCO published a volume entitled *Current Trends in Scientific Research* by the consultant Pierre Auger. The first section of this authoritative work gives a concise account of the development of mathematics in the first half of the 20th Century. Auger surveys the links between mathematics and other sciences and technologies. He draws attention to a "trend towards duality in the training of research workers and technicians" which he defines as "a tendency to separate, in some degree, the theoretical and abstract from the practical and concrete". He then confirms that "the relative importance of theory, and of its most extreme form—mathematics—is undoubtedly increasing as time goes on".

The pattern of research has been considerably affected by the use of electronic computers, and mathematicians are employed at all levels of computer work. Highly trained mathematicians contribute to the actual design and development of the machines. Computers may, in fact, be said to be the result of a happy marriage between mathematics and electronics. Whilst programming work does not necessarily demand advanced mathematical training, the basic formulation of problems to be fed to the machines certainly does.

If computers provide a service to mathematicians by performing calculations which might otherwise be impracticable, they have also created a demand for suitably qualified personnel. At the present time they absorb mathematicians at a rate second only to that of teaching.

Employers in the field of research are listed in such publications as *Industrial Research Laboratories of the United States* (National Academy of Sciences—National Research Council, Washington, D.C., 11th edition 1960). This covers non-governmental laboratories which are devoted to fundamental and applied research. Some 44 entries are listed under Mathematics, and 23 under Statistical Analysis. Many of the other organizations listed do, however, also employ mathematicians. An index to research and administrative officials is appended. As far as government work is concerned, the U.S. Department of Defense is easily the largest employer, though most agencies employ some mathematicians. High figures are, for example, reported by the National Aeronautics and Space Agency, and the Department of Commerce of which the National Bureau of Standards forms part. For those primarily interested in basic research, N.A.S.A. offers the most promising prospects.

Information of this kind is available for other countries. A comprehensive British work is *Industrial Research in Britain*, published by Harrap Research Publications, London. The 4th edition appeared in 1962.

B

Statistics

Mathematical statisticians are required in all branches of science and technology, in government departments, and in business concerns. The demand for them grows as new applications for their work are explored. Business management, quality control, biological research and agriculture are examples of areas in which the use of statistical techniques is currently increasing. It is obviously necessary for the statistician to acquire a working knowledge of each particular subject to which he applies his skills. He must work in close collaboration with subject experts in other fields and must therefore be able to understand their language and their problems. This demands the quality of adaptability and, equally important, a high degree of competence in using bibliographical tools.

Training in statistics requires a good level of mathematical knowledge as a prerequisite. In Britain, only the University of London offers a degree course in statistics. The subject is, however, included in the degree courses for mathematics and economics in other universities. Postgraduate diploma courses are available in such universities as Aberdeen, Birmingham, Cambridge, Manchester, Oxford and Wales (Swansea). Details are given in the calendars of the respective institutions. The Association of Incorporated Statisticians (55 Park Lane, London, W.1) holds examinations at two levels: intermediate and final.

In 1959, the Royal Statistical Society published a 10-page booklet entitled *The Career of Statistician*.

United States universities which offer a master's degree in statistics are listed in the American Mathematical Society's annual survey of *Assistantships and Fellowships in Mathematics*. In April 1962 the American Statistical Association and the Institute of Mathematical Statistics issued their joint booklet called *Careers in Statistics*. It describes what statisticians do by means of six individual case histories, and includes useful sections on education and salaries. The American Statistical Association itself "facilitates the placement of its members through special arrangements with the United States Employment Service and by other means, including the numerous professional contacts of its members".

Operational Research

The role of the mathematician in O.R. is as a member of a team. The team comprises members having a variety of basic skills. This is reflected in the pre-entry requirements to diploma and degree courses in the subject. The Imperial College of Science and Technology (University of London) stipulates, for example, that "Prospective entrants should possess an engineering or science degree or equivalent, together with some industrial experience". The diploma course offered by the London School of Economics and Political Science "is open to men and women who hold a university degree in *any* subject, but a knowledge of Mathematics, Statistics and Economics to the level of the Part I examination of the B.Sc.(Econ.) degree will be assumed".

The methods used in O.R. are largely based on advanced mathematical techniques. The mathematical background to the theory of games, linear programming, and the theory of queues are important facets of the work.

More and more large companies are establishing O.R. departments. These departments vary greatly in size, but it is significant that new departments which have proved their effectiveness tend to grow quite rapidly. Not every industrial company or commercial undertaking is large enough to warrant the creation of its own O.R. department. Operational research services are, however, offered by research organizations and consultants. Attractive openings exist here for mathematicians, and those who decide to specialize in this work still have an opportunity of "getting in on the ground floor".

Salaries are relatively high both for new entrants and those with experience. In the Mathematical Sciences Section of the National Register of Scientific and Technical Personnel, practitioners of O.R. are grouped in the subfield "Mathematics of resource use" which has a high percentage of non-academic employment and the highest median salary of all. Whilst this group also includes the highly paid actuarial profession, the fact that its salary rating is equally representative of the O.R. field is borne out by advertisements in the professional press.

To meet the demand for organized instruction in O.R., new courses are being offered by colleges and universities, and there is in consequence a growing need for lecturers in the field. Operational research societies such as the Operational Research Society (London) undertake to arrange for suitably qualified lecturers to provide on-the-job instruction within industry, and can give advice on opportunities for work of this kind.

Formal training in O.R. at recognized educational institutions is normally given at post-graduate level. The validity of the subject as an academic discipline has now been recognized. The number of courses offered on a regular basis is growing rapidly. In Britain, the courses are usually of one year's duration, except in the case of the London School of Economics and Political Science, where normally a full-time course of instruction extending over one academic year is followed by a further year spent in practical work. The first British course was that introduced by the University of Birmingham in 1958, leading to the Degree of Master of Science in Operational Research. Other full-time courses are offered by the College of Aeronautics (Cranfield), Hull University, the Imperial College of Science and Technology (University of London), and the Welsh College of Advanced Technology, Cardiff.

In 1959, the Operations Research Society of America issued a list of courses in a publication entitled *Formal Offerings in Operations Research*, which it supplements regularly in its periodical publications.

The breadth of subject coverage in courses of instruction in O.R. and cognate techniques may be exemplified by listing the main areas of the syllabus of a typical course. They are statistics, scientific method, industrial psychology, industrial economics, management accounting, data processing, stochastic processes, mathematical programming, planning and control of production and inventories.

Apart from full-time courses, many short courses are organized by various bodies. They are usually concerned with specific O.R. techniques, and the best way of keeping informed about them is to consult the O.R. periodical literature listed in Chapter 12. Furthermore, the national societies are always able to provide such information upon request.

Practical experience is an essential part of O.R. training. This is recognized by those organizations which utilize its techniques, and many of them engage young graduates in appropriate subjects and offer them training within working groups before financing their attendance at academic courses.

Actuarial Science

Actuaries are among the highest paid mathematicians. They belong to a relatively small profession. In Great Britain in 1961 there were less than 1000 Fellows of the Institute of Actuaries. The training is exacting, but the rewards are stimulating. The primary work of an actuary is concerned with long-term financial contracts, particularly in the field of life insurance. It involves, among other things, the balancing of assets and liabilities, the rating of premiums, and the investment of accrued funds. It is responsible work demanding integrity of a high order, and from a mathematical point of view, a thorough grounding in probability theory and statistics, and their practical application. Actuaries are extending their sphere of operation. They have, for some time, found employment in government service. For example, in 1917 the post of Government Actuary was created in Britain. This was followed by the foundation of the Government Actuary's Department, which was intended to deal with national social security activities. Local government and private pension schemes similarly require the participation of actuaries. Statistical activities in which actuaries find work include vital statistics, the construction of life tables, and demography. Stock exchanges are another source of employment. But it is in commerce that opportunities are most rapidly increasing. Actuarial skills are required in the manipulation of investments in large organizations.

An analysis of occupations of Fellows and Associates of the Institute of Actuaries (London) indicates that most are employed in assurance offices. The next highest figure is for whole-time consulting practice. Others are in central and local government service, stock exchange, industry and commerce. It is of interest to note that in 1925 there were no entries under the last heading, whilst in 1961 there were 45.

In Britain, the Commonwealth, and South Africa the necessary qualifications are those awarded by the Institute of Actuaries and the Faculty of Actuaries in Scotland. Tuition is organized by the Actuarial Tuition Service which is administered by a Joint Committee of the two bodies. A list of text books and other useful study literature is given in the *Year Book* of the Institute. This list has the advantage of being graded according to the qualification being studied for, ranging from the Preliminary Examination in Mathematics to the Fellowship Examination. Another publication of the Institute is entitled *Institute of Actuaries: the Examinations, Syllabus and Course of Reading*. More general information is given in its 20-page booklet, *The Actuarial Profession*. The Institute does everything possible to ensure a healthy profession. It has set up an Appointments Board with three stated aims: "To provide authoritative guidance to prospective employers of actuaries who seek such guidance; to provide recognized means of bringing suitable vacancies to the notice of members of the Institute; and to provide an opportunity for members who wish to do so, to discuss matters affecting their careers with a senior member of the profession".

In America, the examining body is the Society of Actuaries, 208 South La Salle Street, Chicago 4, Illinois. There are two parts to the Preliminary Actuarial Examinations. The first is the General Mathematics Examination which is based on the first two years of college mathematics; and the second is the Probability and

Statistics Examination. The actual examinations are held at various centres in the United States and Canada. Another American insight into the profession is given in

IMMERWAHR, GEORGE E., "A career as an actuary", *J. College Placement*, **21**, 43 et seq., 1961.

Sources of information in actuarial science are treated separately in Appendix III as they involve subjects beyond the purely mathematical field.

CHAPTER 2

The Organization of Mathematical Information

510: Mathematics, general
510.1: Philosophy of mathematics
510.2: Handbooks and outlines

THE science of mathematics is a product of human intelligence. Its development has been recorded through history in many forms, the most important of which has undoubtedly been the printed word. The invention of printing from movable type in the 15th Century gave man a ready means of recording and communicating his thoughts and ideas, and has contributed significantly to the attainment of the highly sophisticated mathematical techniques of today.

In this process of communication by means of printed words and other symbols librarians play an active role. Not only do they collect and preserve the vital records, but they analyse and organize them for future use. Libraries are an essential complement to formalized courses of study. An important part of every mathematician's training is that devoted to the efficient use of libraries and the exploitation of all forms of recorded knowledge and sources of information.

Sources of mathematical information may be divided into two main categories: bibliographical and non-bibliographical. Books, periodicals and other printed documents are bibliographical sources. The second category includes professional societies, research organizations, and subject specialists. It should, however, be noted that the *guides* to these do themselves take a bibliographical form. Non-bibliographical sources generally receive too little attention, particularly in relation to their growing importance. The relationship between them and the more conventional bibliographical sources has therefore been made a feature of succeeding chapters.

The need for adequate training in exploiting information sources becomes more strongly felt when formal education ceases. It increases with the degree of specialization one achieves. This is particularly true in today's pattern of scientific research where specialization often takes the form of interdisciplinary synthesis, producing such subjects as mathematical chemistry, mathematical economics, and mathematical linguistics. Whilst, therefore, it is important to know in detail the sources in one's own particular subject field, it is equally important to acquire an understanding of the basic principles of information techniques.

The first essential is a good working knowledge of libraries: what they contain, and the services they provide.

Guides to Using Libraries

A useful introductory work is Jean Kay Gates' *Guide to the Use of Books and Libraries*, McGraw-Hill, New York, 1962. Written by an American librarian, it deals especially with college libraries and their use. It is a textbook for students covering the different kinds of library materials and how they are organized. Ella V. Aldrich's *Using Books and Libraries*, Prentice-Hall, has gone through several editions since 1940. It was originally written for the freshman course in library use at Louisiana State University.

The series of which the present work forms part guides enquirers in all the major subject fields through the relevant literature and and other sources of information. The first volume is the compact but comprehensive *How to Find Out* by George Chandler, Pergamon Press, Oxford, 1963.

Guides to Libraries and Special Collections

The mathematician needs to use the resources of many different kinds of libraries. From student days onwards the university ro college library and the public library will be his standby. On joining the appropriate professional society he will become aware of the resources of its library and information services. Should he find employment in an industrial firm he will in all probability discover that it has an efficient specialized library, for industrial libraries form one of the most rapidly developing segments of the library profession. Particularly if he has recourse to official statistical publications, he will come into contact, either directly or indirectly, with the libraries of government departments. The more specialized his work becomes the greater will be his need to know which libraries further afield will be likely to supply the information he requires.

There are national directories in existence for most countries. The *American Library Directory*, Bowker, New York, 23rd edition, 1962, gives information on nearly 15,000 libraries in the United States and Canada. *Specialized Science Information Services in the United States*, National Science Foundation, 1961, is a directory of information centres which indicates subject specializations, services provided, user qualifications, and publications. Another Bowker compilation, now in its second edition, 1961, is *Subject Collections: a Guide to Special Book Collections and Subject Emphases as Reported by University, College, Public, and Special Libraries in the United States and Canada*. Prepared by Lee Ash, it contains about 20,000 entries arranged in alphabetical order of subjects.

In Britain, the two-volume *Aslib Directory*, London, 1957, provides particulars of 3303 libraries arranged geographically, with a complete index to their subject specialities. Between 30 and 40 libraries are reported as having mathematical collections. A more recent British guide is *Special Library and Information Services in the United Kingdom*, edited by J. Burkett and published by the Library Association in 1961. Separate chapters deal with the different types of libraries, and an appendix by Dr. H. Coblans deals with those of international organizations.

The libraries of professional societies are mentioned in a later chapter, but it is appropriate to quote here a publication of a professional society for documentation, the International Federation for Documentation. Its *Bibliography of Directories of Sources of Information*, The Hague, 1960, lists national and international guides to sources of scientific and technical information in no less than 35 countries. The European Productivity Agency's *International Guide to European Sources of Technical Information*, 1957, first lists national technical information centres "through whom all enquiries should be channelled", and then describes the individual organizations under their respective country headings. There are indexes of subjects and organizations.

Another international guide is Collison, Robert L. *Bibliographical Services Throughout the World, 1950–59*, UNESCO, Paris, 1961. Part 1 records bibliographical activities in various countries and territories. The entries are arranged geographically and include details of any national bibliographies which exist. Part 2 is concerned with the bibliographical activities of international organizations. On pp. 25–30 is a useful list of national corresponding members, again arranged by countries.

Inter-Library Co-operation

No library is ever complete. Even the great libraries of the world such as the British Museum, the Library of Congress, and the Lenin Library have their limitations. But librarians have for many years sought ways to increase the accessibility of recorded information, largely through formal and informal schemes of interlending. These are organized at local, national, and international levels. Every public library in Great Britain, for example, is linked through Regional Bureaux to the National Central Library which maintains vast union catalogues of library stocks. A request for the loan of a particular book not available in a local library will if necessary be circulated throughout the country until a library is found which has a copy and is prepared to lend it. Schemes with similar aims exist in other countries, the majority of which participate in international lending projects. Such activities form the subject of a recent *Guide to Union Catalogues and International*

Loan Centres by L. Brummel and E. Egger, Nijhoff, The Hague, 1961, which contains an article detailing the rules governing international loans and the use of the international loan form. New developments are reported in the *UNESCO Bulletin for Libraries*.

Photocopying and Microcopying

It is often advantageous, both for the requesting library and the lending library, for loan requests to be met by a photocopy of the work in question. Apart from reducing postage costs, this method also obviates the risk of loss of the originals or their damage in transit. Furthermore, the individual enquirer is usually able to borrow such a photocopy for a longer period, or may in some cases be made a gift of it.

There are several ways in which documents are reproduced photographically. Apart from original-size copies there is a variety of microforms including continuous *microfilm*, opaque *microcards*, and their transparent counterparts called *microfiches*. A complete technical report can be reproduced on a single microcard, or a whole book on a roll of microfilm. It is, of course, necessary for this material to be used in a reading machine which throws up an enlarged image onto a screen. The production of photocopies is governed by copyright laws, but many publishers subscribe to so-called *fair copying declarations* which permit copying in compliance with specified conditions. The *Directory of Photocopying and Microcopying Services* published by the International Federation for Documentation, 2nd edition, 1955, lists 200 reproduction services in 58 countries with detailed descriptions of the services offered and prices charged. Also included are the texts of the British and United States fair copying declarations.

An increasing number of books and periodicals are being produced on microfilm to meet demands for out-of-print publications, and for the convenience of libraries wishing to conserve their limited shelf accommodation. A complete file of the *Journal of the Institute of Actuaries* for 1850–1959, for example, is available on microfilm from Micro Methods Ltd., England. Pioneers of this type of reproduction are University Microfilms of Ann Arbor, Michigan. An international list of micro-publishers is given in *UNESCO Bulletin for Libraries*, **16**, 198–205, July–Aug. 1962.

Library Catalogues

The function of the catalogue is to make the contents of the library known, to indicate the location of every item within the library. Catalogues take different forms. They may be printed books, sequences of filed index cards, loose-leaf sheets locked in binders, visible index strips, or even punched cards. Classic examples of printed catalogues are those of the British Museum and the U.S. Library of Congress. Most public, university and college libraries have adopted index cards as being the most satisfactory from the point of view of updating. The Library of Congress printed catalogue was in fact produced from arrays of cards laid down in page form suitable for photo-mechanical reproduction. Catalogue cards produced by the Library of Congress may be purchased for use by other libraries. Each card bears full particulars of the book in question and the appropriate classification symbols both from the Library of Congress Classification and the Dewey Decimal Classification. Many new books indicate the relevant card numbers to facilitate ordering.

Whatever the form a catalogue takes, its entries will always follow a similar pattern. Minimum details are author, title, date and location symbol. The publisher and date of publication are also normally included. Generally the location symbol coincides with the classification notation since most libraries arrange their books on the shelves in a classified order. There are many rules governing the cataloguing of books which are designed to ensure consistency of treatment.

It is essential for the mathematician to familiarize himself with the arrangement of the library catalogues he intends to use. Much will be missed if too much reliance is placed on the shelf arrangement of the books themselves. This would be rather like consulting the alphabetic arrangement of articles in an encyclopedia and ignoring the analytical index. Most well-organized libraries issue a guide to the use of their catalogues.

Three aspects of a book may be used as the basis for arranging catalogue entries: its author, its title, and its subject. In consulting indexes and catalogues care should be taken to understand the filing rules that have been followed. For example, there is the difference between letter-by-letter and word-by-word arrangements. The former places *Logarithms* before *Log cabins*, whilst the latter places *New York* before *Newspapers*. Abbreviations, including numbers, are filed as if they were written out in full: M' and Mc equal Mac, St. equals Saint, 8 equals Eight.

When author, title, and subject entries are interfiled to form one alphabetical arrangement the result is known as a *dictionary catalogue*. Cross references are included to guide the user. It can, of course, happen that the same filing word may be the name of an author, a place name, or a book title. The word "Bedford" is an example. In such cases the accepted filing order is:

Persons—authors	e.g. Bedford, John Harold
—subjects	Bedford, John Harold (a biography)
Places —authors	Bedford Public Library
—subjects	Bedford (a town guide)
Titles	*Bedford Parks*

A *classified catalogue* is one in which the subject entries are filed separately in the order of classification notations, thereby paralleling the sequence of books on the shelves. This requires a subject index to enable the user to convert the name of his subject into the appropriate notation. The author index, including titles, is also in a separate run.

The library catalogues of professional societies and other organizations are often published in book form. The *Register of Mathematical Publications in N.S.A.*, compiled by Jean Chandler Andrews, indicates the holdings of the U.S. National Security Agency Library. It was published in 1954 and contains 133 pages.

Other examples are given in later chapters. Catalogues of this kind are particularly helpful in locating highly specialized material.

Classification of Mathematics

The object of classification in libraries is to bring together like material. Whilst printed records appear in different *forms* (books, periodicals, reports, etc.) they are classified by *subject*, since this is the approach most used by enquirers. Whilst new systems of classification are constantly being evolved, particularly to meet the requirements of those developing the use of electronic computers for information storage and retrieval, most libraries have adopted one or other of the established sets of published schedules. Undoubtedly the scheme most likely to be met with is the Dewey Decimal Classification, which was first published in the United States towards the end of the last century and is now in its sixteenth edition. It, and its descendant the Universal Decimal Classification, are the most widely used in libraries throughout the world.

Basically, the Dewey Decimal Classification consists of a distribution of the nine major divisions of human knowledge, plus a generalia class, through a three-figure array of main classes, as follows:

000: Generalia	500: Pure Science
100: Philosophy	600: Applied Science
200: Religion	700: Fine Arts
300: Social Sciences	800: Literature
400: Language	900: History and Geography

Each main class is subdivided into ten subclasses:

500: Pure Science	520: Astronomy
510: Mathematics	530: Physics, etc.

Further divisions are made in a similar way, each time decreasing the extension of the subject specified:

510: Mathematics	512: Algebra
511: Arithmetic	513: Geometry, etc.

The process continues after the decimal point:

517: Calculus

517.1: Infinitesimal calculus   517.3: Integral calculus
517.2: Differential calculus   517.4: Calculus of variations

The extreme simplicity of this method of subdivision, plus the fact that Arabic numerals are almost universally understood, has made the Decimal Classification an important instrument in the documentation of recorded information at an international level. Many articles in technical periodicals bear a Decimal Classification symbol; bibliographies and abstracting publications are arranged by the Decimal notation. Thus a mathematician knowing the symbols for his own field of specialization can readily exploit foreign sources which might otherwise remain inaccessible. He is immediately at home in a strange library.

Simple as the decimal notation is to memorize, additional mnemonic devices have been built into the scheme in the form of common subdivisions which are applicable throughout the schedules. Some examples are seen in the arrangement of the chapters which follow. 03 indicates dictionaries, 05 periodicals, and 09 history when added to subject numbers. When the basic notation ends in 0 this figure need not be repeated. Thus 510.3 equals mathematical dictionaries, 510.5 equals mathematical periodicals, and 510.9 equals history of mathematics.

This very brief outline of the scheme can be supplemented by the *Guide to the Use of Dewey Decimal Classification*, Forest Press, Lake Placid, N.Y., 1962, which is based on the practice of the Decimal Classification Office at the Library of Congress.

Philosophy and Foundations of Mathematics

In the preface to his book *Mathematics Manual: Methods and Principles of Mathematics for Reference, Problem Solving, and Review*, McGraw-Hill, New York, 1962, Frederick S. Merritt talks of "The vine of mathematics, rooted in logic". A study of logic is indeed essential to a full understanding of the nature of mathematics. Quite naturally, the Dewey Decimal Classification makes provision for the subject in its philosophy class 100. Works on symbolic logic are specifically catered for in subclass 164. Whilst the primary function of classification is to bring like things together, it is inevitable that some subjects which in practice are *studied* together will be separated in the classification schedules. In Chapter 11 attention is drawn to other cases in which subjects of interest to the mathematician are classified outside the 510 heading.

An excellent "Bibliography of symbolic logic, 1666–1935" by Alonso Church appeared in Volume 1 of the *Journal of Symbolic Logic*, 1936. Additions and corrections were published in a later volume. This journal is the official organ of the Association for Symbolic Logic. It contains papers on symbolic logic and studies in the history of logic, and regularly features an extensive review section of relevant literature. Another important journal in the field is the German *Zeitschrift für mathematische Logik und Grundlagen der Mathematik*, Berlin, which commenced publication in 1955.

A Survey of Symbolic Logic by Clarence Irving Lewis, Dover, New York, 1960, also includes a very detailed bibliography. This edition is actually an adapted republication of the original work issued in 1918 by the University of California Press. In association with Cooper Harold Langford, the same author also produced *Symbolic Logic*, whose second edition was published by Dover in 1959.

An examination of mathematical books in a library will reveal that certain publishers tend to specialize in books on a particular subject. These are often produced as a uniform series. The North-Holland Publishing Company of Amsterdam is currently issuing such a series entitled *Studies in Logic and the Foundations of Mathematics*, edited by L. E. J. Brouwer, E. W. Beth, and A. Heyting. Some two dozen titles have already appeared. Representative of these are Beth, E. W., *The Foundations of Mathematics*, and Fraenkel, Abraham A., *Abstract Set Theory*, which again contains a useful bibliography. Van Nostrand's *University Series in Undergraduate Mathematics* contains additional works

c

on set theory including Patrick Suppes' *Axiomatic Set Theory*. His *Introduction to Logic* is also in this series.

There are several great names which will always be linked with the development of the foundations of mathematics. Bertrand Russell is prominent among these. His *Principles of Mathematics* was originally published in 1903 by the University Press, Cambridge, and later published with a revised introduction. With A. N. Whitehead he produced the extremely important *Principia Mathematica*, 3 volumes, Cambridge University Press, 1910–13. Alfred Tarski's contributions to mathematical logic and meta-mathematics were published by Oxford University Press in 1956 under the title *Logic, Semantics, Metamathematics*. The papers this volume comprises were all originally published in the years 1923–38 in various languages. Numerous references guide the reader to related writings. *Boolean algebra* and *Hilbert spaces* are terms well known to mathematicians. Both George Boole and David Hilbert made important contributions to the philosophy of mathematics. *Studies in Logic and Probability*, Waters, London, 1952, contains Boole's *Mathematical Analysis of Logic* and other writings. With P. Bernays, Hilbert wrote *Grundlagen der Mathematik*, 2 volumes, Springer, Berlin, 1934–9.

Three more recent works which should be noted are:

FREUND, JOHN, *A Modern Introduction to Mathematics*, Prentice Hall, New York, 1956.

ROSSER, JOHN, *Logic for Mathematicians*, McGraw-Hill, New York, 1953.

WILDER, RAYMOND L., *Introduction to the Foundations of Mathematics*, Wiley, New York, 1952.

Exercises

1. List some of the ways in which a librarian can help a student of mathematics.
2. Why is it more important for a mathematician to know how to use a library than to memorize a long list of books on his subject?
3. Distinguish between bibliographical and non-bibliographical sources of information and discuss their relative importance.
4. Write notes on the Dewey Decimal Classification with particular reference to mathematics.

CHAPTER 3

Mathematical Dictionaries, Encyclopedias and Theses

510.3: Dictionaries and encyclopedias
510.4: Essays and lectures

THE scope of this chapter heading has been widened a little to include something about translations of foreign mathematical literature. At the same time it has been restricted to the extent that Russian mathematical literature is dealt with separately in Appendix I.

Dictionaries

General Dictionaries

Though the concern here is primarily with dictionaries of mathematics, it should be noted that the better *general* dictionaries have a much wider use than the simple verification of spellings and definitions. Through their careful distinctions in shades of meaning, their explanations of historical derivations, and their cross references, they can be a considerable aid in initial orientation in unfamiliar and quite complex subjects. The following should be examined:

Oxford English Dictionary, edited by Sir James Murray *et al.*
Webster's New International Dictionary
FUNK and WAGNALL's *New Standard Dictionary*

Mathematical Dictionaries

An important work is the *Mathematics Dictionary* edited by Glenn and Robert C. James, Van Nostrand, 1959. It is described in the preface as "by no means a mere word dictionary, neither is it an encyclopedia. It is rather a correlated condensation of mathematical concepts, designed for time-saving reference work. Nevertheless the general reader can come to an understanding of concepts in which he has not been schooled by looking up the unfamiliar terms in the definition at hand and following his procedure to familiar concepts." This statement is worth re-reading, for this is the way a dictionary should be used. An appendix contains a selection of tables and a useful list of mathematical symbols, and there are indexes in French, German, Russian and Spanish.

There are two other recent dictionaries which may be noted here. Less expensive than James is C. C. T. Baker's *Dictionary of Mathematics*, published by Newnes, London, in 1961. It is a concise work "suitable up to degree standard". A cheaper work still, and one which was "written specially for young people", is *A Dictionary of Mathematics* by C. H. McDowell, Cassell, London, 1961.

Selection of a dictionary must be based on the requirements of the individual user. The only real way to test the suitability of a particular volume is to put it to practical use. Two or three should be available in most libraries. It is to be expected that the less expensive works will be more selective in their coverage. Baker, for example, does not include *topology*, but limitations of this sort must be matched against the price and the educational level for which the dictionary was compiled.

Other dictionaries of interest to the mathematician occur in more specialized fields, and are described in the appropriate chapters. Representative of these is the *Dictionary of Statistical Terms* by M. G. Kendall and W. R. Buckland published by Oliver and Boyd, Edinburgh, 2nd edition, 1960.

For those with a knowledge of foreign mathematical terminology, interlingual dictionaries can also be used to advantage in clarifying the meanings of terms in one's native tongue. Some German terms, for example, will be found to be more expressive than their English counterparts.

Translations

Mathematics is international. New developments are reported in the scientific press of many countries, and in many different languages. Abstracting and translating services endeavour to facilitate the use of foreign-language literature, but it is frequently necessary for the mathematician to be able to read material in its original form. It is for this reason that many universities have a language requirement in their degree programmes. The three most useful languages are French, German and Russian. A reading knowledge of *two* foreign languages is often specified for doctoral degrees.

It is only natural that English-speaking mathematicians will first try to discover whether a particular foreign work has appeared in translation. In this connection the American Mathematical Society's *Notices* gives details of Russian books currently being translated into English by the Society, under grants from the National Science Foundation, under the heading "Mathematics in translation". This feature also reports translations, both from Russian and from other languages, being prepared by other organizations.

A standard reference work on translations is the *Index Translationum: International Bibliography of Translations* published in Paris by UNESCO. The stated object of UNESCO in publishing the index is to provide an annual listing of all translated books published throughout the world on any subject. Some idea of its proportions may be gained from the fact that the 1960 annual volume contains more than 31,000 items representing 58 different countries. The section on the United States is compiled by the General Reference and Bibliography Division of the Library of Congress, and that on the United Kingdom by the British Museum. Translations are first arranged according to the country in which they were published, and are further subdivided by the 10 major headings of the Universal Decimal Classification. Thus mathematics comes under the heading "Natural and Exact Sciences". A complete alphabetical index of authors is provided.

The system of the International Organization for Standardization for transliterating Cyrillic alphabets (e.g. Russian) is used. Each entry follows the plan: author; title of the translation; place of publication; name of publisher; number of pages, etc.; price (in the country of translation); the language in which the work was written, or from which it was translated; title of the original work; place, publisher and date of the original.

A recent *Survey of Translation Activities in Universities, Societies and Industry in the Fields of Science and Technology*, prepared by the Special Libraries Association, New York, indicates that the U.S. sources most frequently checked for existing *unpublished* translations are the S.L.A. Translations Center in Chicago, and the Office of Technical Services. In Britain, Aslib is the principal source. Its 1962 publication, *The Foreign Language Barrier*, edited by C. W. Hanson, deals with the learning of languages by scientists, cover-to-cover translations, and pools and services.

Interlingual Mathematical Dictionaries

It will be remembered that the *Mathematics Dictionary* of Glenn and Robert C. James has indexes in French, German, Russian and Spanish. Indexes in the same four languages are given in the large *International Dictionary of Applied Mathematics*, which was compiled by experts in different fields under the general editorship of W. F. Freiberger, and published by Van Nostrand, Princeton, N.J., in 1960. Dictionaries which cover more than two languages are called *polyglot*.

The majority of interlingual dictionaries are restricted to two languages. These are termed *bilingual*. They are usually more detailed than polyglot dictionaries, often giving individual terms in the context of phrases. In mathematics, German and Russian have received most attention. German–English compilations include:

> HERLAND, LEO., *Dictionary of Mathematical Sciences: German–English and English–German*, 2 volumes, Ungar, New York, 1951–4.
> The statistical entries are by Gregor Sebba and the commercial entries

by Robert Grossbard. Numerous cross references increase the value of
the work.

HYMAN, C., *German–English Mathematics Dictionary*, Interlanguage
Dictionaries, New York, 1960.

Whilst lacking some of Herland's helpful features, it is very useful
when dealing with the newer terminology.

KLAFTEN, E. B., *Mathematical Vocabulary, English–German/German–
English*. Wila Verlag, Munich, 1961.

Has an unusual arrangement in that each part is divided into 8 sections:
general terms, arithmetic, algebra, plane geometry, solid geometry,
trigonometry, calculus, and co-ordinates. Of handy format, it gives
examples of usage as well as individual terms.

A compact and inexpensive work published in 1959 by the
VEB Deutscher Verlag der Wissenschaften, Berlin, has two
complementary title pages to explain its purpose: *Mathematical
Dictionary—Russian–English, with a Short Grammar*, and
*Mathematisches Wörterbuch—Russisch–Deutsch, mit einer kurzen
Grammatik*. An outline Russian grammar in both English and
German precedes the trilingual vocabulary. The Russian words
determine the alphabetical order, and each is followed by German
and English equivalents. Some names of mathematicians are given,
together with very brief biographical notes. These appear in the
German column but not in the English.

Yet another function is filled by the *German–English Mathematical Vocabulary*, by Sheila Macintyre and Edith Witte, published by Oliver and Boyd, Edinburgh, 1956. An inexpensive
work of less than 100 small pages, it comprises a vocabulary and
grammatical sketch. It is essentially a practical tool for those who
need an introduction to the language with a mathematical bias.

Although Russian is separately dealt with in Appendix I, it is
appropriate to mention here an excellent dictionary which was
prepared under the auspices of the American Mathematical
Society, the National Academy of Sciences of the U.S.A., and the
Academy of Sciences of the U.S.S.R. It is the *Russian–English
Dictionary of the Mathematical Sciences*, compiled and edited by
A. J. Lohwater with the collaboration of S. H. Gould, published
by the Society in 1961. It contains over 15,000 terms preceded by a
short grammar of the Russian language.

Specialized dictionaries are also available for other languages. Italian, for example, has Giorgio Guidi's *Prontuario Tecnico Inglese–Italiano, Raccolta di Termini Tecnici dell' Uso Corrente della Lingua Inglese Relativi alla Matematica, ecc*, Rome, 1947.

Guides to Foreign Scientific and Technical Dictionaries

Many libraries and most individuals content themselves with acquiring a selection of general scientific and technical dictionaries in the more popular languages, and these are often sufficient in translating mathematical texts. The *German–English and English–German Dictionary for Scientists* by O. W. and Irma S. Leibiger (Edwards, Ann Arbor, Mich., 1950), for example, certainly includes a large number of mathematical terms.

It is, however, necessary, on occasions, to be able to identify more specialized works, particularly in the less common languages. Two compilations will prove of value here:

> LIBRARY OF CONGRESS. *Foreign Language–English Dictionaries*, Volume 1: *Special Subject Dictionaries, with Emphasis on Science and Technology*, 1955.
>
> Entries are arranged alphabetically by subject. Within each subject bilingual dictionaries, in an alphabetical sequence of languages, are followed by polyglot dictionaries.
>
> UNESCO. *Bibliography of Interlingual Scientific and Technical Dictionaries*, Paris, 4th edition, 1961.
>
> Entries are arranged according to the Universal Decimal Classification, and there are English, French and Spanish indexes of authors and subjects.

An attempt to keep up-to-date by means of loose-leaf publication is being made by Karl Otto von Saur's *Technik und Wirtschaft in Fremden Sprachen*, Verlag Dokumentationen der Technik, Munich. This is an international bibliography of subject dictionaries and encyclopedias.

Encyclopedias

Encyclopedias are often either overlooked or spurned by those seeking scientific and technical information. This is due to two main reasons. Firstly, general encyclopedias cover the entire range of human knowledge, and are therefore sometimes considered to treat their material superficially. Secondly, they take so long to compile, edit and print that they are out of date long before they are published.

To put these arguments into proper perspective it is necessary to examine the functions of the general encyclopedia. These are twofold. By the very wealth of its factual content it acts as a comprehensive fact-finding tool which yields an immediate end-product. But it is not necessarily intended as an end in itself. An equally important function is that of orientating the enquirer.

The articles which comprise a first-rate encyclopedia are prepared by specialists. They write according to a preconceived plan which is designed to relate the contributions on individual topics to the context of knowledge in general. Cross references between articles lead the enquirer to related topics. Each article contains a bibliographical guide to other more specialized or detailed writings on the subject with which it deals. It also gives leads to further research in the form of terminology which can be noted as used as key-words; in establishing date limits; in providing the names of those who have made important contributions to the subject in question; and in specifying the countries and institutions in which research has been carried out. These features make the general encyclopedia a very valuable starting point for information searches which may ultimately end with the examination of a highly abstruse monograph, or even the writing of a letter to a research institute.

On the question of up-to-dateness, two points should be noted. The first concerns their method of revision. Broadly speaking, there are two ways in which general encyclopedias are revised. They are either completely revised in every section and then published as new editions, or else they undergo what is known as *continuous revision*. This means that editorial staff are constantly working at the task of keeping the encyclopedia up-to-date, introducing new facts, recording historical events, replacing outdated illustrations. Whilst in the case of an encyclopedia having completely revised editions such editions will only be published at intervals of five years or more, printings of the continuous revision encyclopedia may be made annually or even oftener. At each successive printing amendments made since the last printing are incorporated. This results in a certain unevenness of revision, but is a method adopted by several of the more important works. Updating is also achieved in several cases by the publication of yearbooks.

The second consideration in the question of up-to-dateness is that mere age by no means invalidates all information. The history of so-called invention is full of cases in which the development of a discovery made many years previously had to await advances in another field to make it practicable. The principle of jet propulsion was not discovered by Whittle, though it was he who developed its practical application.

The publishers of important encyclopedias usually operate a research service, and draw upon wide specialist resources. They are often able to put enquirers into touch with sources of information. Reprints of encyclopedia articles are sometimes produced, and they make useful monographs. An example of such services is the series of *Home Study Guides* produced by the Library Research Service of the *Encyclopaedia Britannica*. One volume in the series is entitled *Mathematics*. It is intended to guide the reader through the mathematical contents of the multi-volume encyclopedia.

The evaluation of encyclopedias is adequately dealt with in Constance M. Winchell's *Guide to Reference Books*, American Library Association, Chicago, 7th edition, 1951.

General Encyclopedias

It is generally advisable to consult several different encyclopedias as the treatment of individual subjects varies, sometimes quite considerably, from one to another.

The following are four of the most important English-language general encyclopedias:

Chambers's Encyclopaedia. Newnes, London, 1959.
  This appears in 15 volumes, and does not follow the policy of continuous revision. It is anticipated that completely new editions will appear every 4 or 5 years.
Collier's Encyclopedia. Collier, New York, 2nd edition, 1962.
  This completely revised edition is in 24 volumes, 4 more than the 1st edition. The last volume contains 11,500 bibliography entries and the index.
Encyclopedia Americana. New York and Chicago, Annual printings.
  Uses continuous revision and the *Americana Annual* (1923–) to keep up-to-date. Most of the articles are short. Volume 30 contains the index, which has an alphabetical, dictionary arrangement and is revised with each printing.
Encyclopaedia Britannica. Chicago, 14th edition, Annual printings.
  Still of great value where up-to-dateness is not essential is the 9th edition (1875–89) in 25 volumes. In the current edition the previous monograph-type articles have been replaced by shorter articles. The index, which should always be consulted when using the Encyclopaedia, is in Volume 24. Continuous revision and the *Britannica Book of the Year* (1938–) are its means of keeping up-to-date.

Mathematical Encyclopedias

Encyclopedias are produced which cater for more specialized needs. Scientific and technical encyclopedias vary in the degree of their specialization, some covering a number of fields, others restricted to an individual discipline. In their respective areas they offer more detailed information than the corresponding sections of general encyclopedias, and their bibliographies may be more ample.

An example of a multi-volume special encyclopedia is the *McGraw-Hill Encyclopedia of Science and Technology*, New York, 1960. The basic 15 volumes are subject to continuous revision and are supplemented by year books, beginning in 1962. Most of the longer articles are followed by bibliographies. *Van Nostrand's Scientific Encyclopedia* is shorter, but also of great value.

A mathematical encyclopedia which contains bibliographies is the large *Encyklopädie der mathematischen Wissenschaften mit Einschluss ihrer Anwendungen*, Teubner, Leipzig, 1898/1904–1904/35. Despite its age it remains a valuable source of reference. A new edition did in fact commence publication in 1950.

Another important German work is the *Encyklopädie der Elementarmathematik*, Deutscher Verlag der Wissenschaften, Berlin, 1954–.

Index Lexicorum, edited by Gert A. Zischka and published by Hafner, Vienna, 1959, is a bibliography of general and special encyclopedias and special dictionaries of all fields published throughout the world since the invention of printing. It lists over 7000 works, arranging them by title in subject groups, and includes a detailed index of authors and keywords.

Theses

Published essays and lectures on mathematics, such as the *Popular Lectures in Mathematics* series of Pergamon Press, are classified at 510.4. With them can conveniently be grouped dissertations or theses. Great strides have been taken in recent years both in making degree theses more readily available and in bringing their existence to the notice of those who can best make use of them. A considerable amount of valuable research work has for long remained unexploited in the files of university archives. Sometimes theses see the light of day in the form of published books, but more often they are shelved in their original form, having once fulfilled their primary academic function. Several organizations, however, are now tackling the problem of thesis documentation. The sources of information for three countries will be described.

In the United States, University Microfilms Inc. of Ann Arbor, Michigan, have for several years published a bibliography of theses entitled *Dissertation Abstracts*. This is "a monthly compilation of abstracts of doctoral dissertations submitted to University Microfilms, Inc. by more than 135 co-operating institutions". Naturally, the listing is only as comprehensive as these institutions make it. The abstracts are quite full, and are grouped under subject headings. Subject and author indexes are included in each issue. Copies of the dissertations are available either as photocopies or as microfilms from the publishers.

While *Dissertation Abstracts* is the principal source of reference, other compilations of more restricted scope have been published from time to time. An example in the mathematics field is Summers, E. G. and Stochl, J. E., "Bibliography of doctoral dissertations completed in elementary and secondary mathematics from 1918–1952", *School Science and Mathematics*, **61**, 323–35, 431–9, 1961.

The standard British work is the *Index to Theses Accepted for Higher Degrees in the Universities of Great Britain and Ireland*, published by Aslib, London. The first volume, covering the 1950–51 academic year, was published in 1953. Volume 10, for 1959–60, records 3195 entries arranged in subject groups. There is an author index and an index to subject headings. In this particular volume the Mathematics heading is further subdivided into Theory of Numbers, Algebra, Geometry, Analysis, Numerical Methods, and Probability and Statistics. A tabulated statement of the availability of theses from each university is provided.

French theses are deposited with the Bibliothèque Nationale and the Bibliothèque de la Sorbonne in Paris. Exchange systems exist between the latter and universities in other countries. Photocopies and microcopies of French theses are usually available from the Service de Documentation of the Centre National de la Recherche Scientifique in Paris. Mathematical listings have been published in *French Doctoral Theses—Sciences* by French Cultural Services of New York. A more comprehensive listing is the *Catalogue des Thèses et Ecrits Académiques*, which is issued in the form of a supplement to *Bibliographie de la France* under the authority of the Ministère de l'Education Nationale.

Duplication of research work is an obvious waste of any country's resources of scientific manpower, a subject which has received a great deal of government attention. Vast sums of money are invested in university research activities, and one way in which the dividends may be increased is by more efficient use of doctoral theses.

Exercises

1. List the features you would look for in assessing the value of a mathematical dictionary for a particular class of user. First define the class of user you have selected.
2. As a mathematician, for what kinds of information would you consult a multi-volume general encyclopedia?
3. Why do you consider that a knowledge of sources of information on theses is important?
4. Write notes on the question of up-to-dateness in encyclopedias.

CHAPTER 4

Mathematical Periodicals and Abstracts

510.5: Mathematical periodicals

As was noted in Chapter 2, periodicals devoted to a particular subject are classified, according to the Dewey Decimal Classification, by adding the figures 05 to the notation of that subject. Information on the periodical literature of such subjects as mathematical statistics and operational research is given in the appropriate chapters. *Applied Statistics*, for example, is mentioned in Chapter 11, and *Mathematics of Computation* in Chapter 7. If the Dewey number for probability theory is 519, then a journal on that subject would be given the notation 519.05. The techniques described in this chapter can be equally well applied to the periodical and abstracting literature of other subjects of interest to the mathematician.

Importance of Periodicals

Every scientific book is out of date even on the day it is published. During the time-lag between the author putting his first words on paper and the appearance of the finished product new opinions, new theories, new facts will have come to light which it is virtually impossible to incorporate. This applies particularly in the case of works treating subjects of current importance and in which active research is being carried out. Books are not, therefore, always the ideal medium for the dissemination of scientific information. The research worker demands a rapid service, and this is provided to a more satisfactory degree by periodicals. It must, however, be confessed that periodicals themselves are being overwhelmed by the pace of today's scientific progress. The American Mathematical Society regularly publishes details of the backlogs which have accumulated in such important publications as the *American Journal of Mathematics, Annals of Mathematics, Canadian Journal of Mathematics*, and the *Proceedings* and *Transactions of the American Mathematical Society*. In some subjects the position is considered so acute as to warrant the use of electronic computers in accelerating the output of urgently needed information. The American Chemical Society's *Chemical Titles*, in which the entries are sorted under key words, is produced almost entirely by computer.

The importance of periodicals is evidenced by the rate at which new titles appear. As research becomes more specialized, so the subject scope of periodicals becomes more restricted.

There is a constant reciprocal stimulus between science and technology. With the development of the newer technologies the demand for more basic research increases. It is for this reason that topology is receiving so much attention today. The wealth of new information produced and the need for a convenient medium for its publication are factors which occasioned the birth, in 1962, of *Topology: an International Journal of Mathematics*, issued quarterly by Pergamon Press.

Guides to Periodicals

The history of scientific periodicals goes back to the 1660's, to the inception of the world-renowned *Philosophical Transactions* and the *Journal des Sçavans*. At first, periodical publications were the product of large academies and for many years remained few and select. Today, some estimates of the number of scientific and technical periodicals reach 100,000. These are published in over 60 languages. An account of the early history of this form of literature is given in:

KRONICK, DAVID, *A History of Scientific and Technical Periodicals: the Origins and Development of the Scientific and Technological Press, 1665–1790*, Scarecrow Press, New York, 1962.

There are some 1500 journals which contain mathematical contributions. The list of periodical titles given in the latest index issue of *Mathematical Reviews* constitutes a comprehensive, international list of periodicals of mathematical interest. Another list of international scope is that in the mathematics section of Malclès, Louise Noëlle, *Les Sources du Travail Bibliographique* (see p. 92).

Ulrich's Periodicals Directory is revised every few years by Bowker, New York. The latest edition is the 10th, 1963. Some 20,000 titles from all over the world are grouped under 200 subject headings. Typical of these are Abstracts and Abstracting Services, Computers, Cybernetics, Mathematics, and Statistics. Up-to-date information on each title includes price; frequency of issue; name and address of publisher; whether the periodical carries abstracts, reviews, or bibliographies; whether it is abstracted or indexed in any other publication; and its date of origin.

A new British work, published by the Library Association, London, in 1962, is a *Guide to Current British Periodicals* edited by Mary Toase. Approximately 3800 titles published in the United Kingdom are arranged according to the Dewey Decimal Classification, and there is a comprehensive index.

Different Kinds of Periodicals

In general terms, mathematical periodicals are issued by three different kinds of publishers: universities, societies, and commercial houses. There are a few periodicals issued by government departments which are of value to the mathematician, but they are not so numerous as in other fields. The *Journal of Research of the National Bureau of Standards; B: Mathematics and Mathematical Physics* is a publication in this category.

Universities

In mathematics, the universities of many countries continue to make valuable contributions by the publication of the results of

D

basic research. The following are examples from the United States:

Duke Mathematical Journal. Duke University Press, Durham, North Carolina, 1935–.

Illinois Journal of Mathematics. University of Illinois Press, Urbana, Ill., 1957–.

Journal of Mathematics and Physics. M.I.T., Cambridge, Mass., 1921–.
Michigan Mathematical Journal. University of Michigan Press, Ann Arbor, Mich., 1952–.

Notre Dame Journal of Formal Logic. University of Notre Dame Press, Notre Dame, Indiana, 1960–.

Pacific Journal of Mathematics. University of California Press, Berkeley, Calif., 1951–.

Annals of Mathematics is edited with the co-operation of Princeton University and the Institute for Advanced Study, and published by Princeton University Press.

Societies

The periodical publications of mathematical societies (proceedings, transactions, etc.) may be classified, with other works relating to societies, under 510.6, which is the subject of Chapter 5. This is reasonable in view of the fact that they report society business, give notice of meetings, and include other matters of interest to members. They are, however, considered here for their subject content. When a particular society has a large publishing programme it may issue several periodicals, each devoted to a different aspect of the society's activities. The American Mathematical Society is a good example. It publishes the results of original research in both its *Proceedings* and its *Transactions*, with longer papers usually in the latter. The *Bulletin* is the Society's official organ, containing reports of meetings, book reviews, the full texts of invited addresses, and other matter of interest to professional mathematicians. Its fourth publication, the *Notices*, contains details of meetings, personal notes and other news items.

The London Mathematical Society was established in 1865, and in the same year commenced publication of its *Proceedings* which now contains more lengthy papers. Shorter papers are printed in the society's *Journal* (1926–) which also contains the records of proceedings at meetings, obituary notices, and other matter of interest to members. Other important British periodicals are the Royal Society of London's *Philosophical Transactions; Series A: Mathematical and Physical Sciences* (1665–), and *Proceedings; Series A: Mathematical and Physical Sciences* (1800–); and the Mathematical Association's *Mathematical Gazette* (1894–), published by Bell, London.

It is not possible to give a full list of mathematical society publications here, but it may be noted that the organizations listed in the next chapter between them issue an important body of periodical literature.

Commercial Publishers

In many subjects, whilst the publication of research is largely the province of institutional periodicals, the practical applications of that research are reported by the commercial press. This is not true to the same extent in mathematics, except perhaps in such specialized cases as electronic computers.

Two important titles of recent origin are:

Advances in Mathematics. Academic Press, 1961–.
Each volume is issued in paper-bound parts at irregular intervals. These are later available as hard-bound books.

Topology, an International Journal of Mathematics. Pergamon Press, 1961–.
A research journal with special emphasis on topology, topological techniques, Lie groups, differential geometry and algebraic geometry.

There are several instances where mathematical societies place the publication of their periodicals in the hands of commercial houses. Bell of London, for example, is the publisher for the Mathematical Association. Such, however, are purely arrangements to suit administrative convenience.

Foreign-language Periodicals

Most countries with an academic tradition produce at least one mathematical periodical which is internationally recognized.

English-language abstracts of their contents regularly appear in *Mathematical Reviews* (see p. 46). In France, for example, Gauthier-Villars of Paris publishes the *Journal de Mathématiques Pures et Appliquées* (1836–). In Germany, Springer of Berlin publishes *Mathematische Annalen* (1869–), and *Mathematische Zeitschrift* (1918–). Scandinavia produces the important *Acta Mathematica* which was begun by Mittag-Leffler in Stockholm in 1882.

Other foreign titles may be traced in *Ulrich's Periodicals Directory*, described above. Some idea of their relative importance may be gained by observing the frequency with which they are cited as references in research papers and analysed in abstracting publications.

Locating and Using Periodicals

Periodicals may be used for current perusal, to keep abreast of new developments, and for retrospective searching, to locate specific pieces of information in past issues. Articles which may have little relevance to one's present interests may assume a greater importance at a later date. By considering some of the work involved in searching back files of periodicals, the value of systematic current perusal, or *scanning*, may be better appreciated.

Searching Periodical Literature

It is the normal practice in libraries to preserve all issues of those periodicals which report information of permanent value. They are usually bound into volumes to facilitate shelving and handling, and to prevent individual issues from becoming damaged or even lost. Each volume will contain an index of authors and subjects produced by the publisher. It may be included in one of the issues or sent separately at the end of the year. The value of periodicals for which no index is produced is considerably reduced. Some publishers, on the other hand, issue cumulative indexes covering perhaps 10 years or more.

One's choice of which files of periodicals to examine will be conditioned by various factors. Naturally, those are examined first which are most readily available and which are known to be likely sources. Most people who are accustomed to periodical searching have their favourites which they always try first. Abstracting and indexing publications, considered in the following sections, provide short-cuts and obviate the tedious job of checking annual indexes of individual titles.

It is essential when undertaking a lengthy search to keep a careful record of the sources which have been examined, and the key words which have been used when consulting indexes. Information discovered while the search is in progress may necessitate a re-examination of the indexes already consulted, this time under one or more new subject headings. It may also happen that a second person is brought in to assist with the search, and he will need to know what ground has already been covered. Search records are conveniently made on index cards and filed in alphabetical order of the titles checked.

When a potentially useful reference is located it should be fully transcribed, preferably in one of the accepted bibliographical forms. A typical reference might read:

> BOOT, C. G., "On trivial and binding constraints in programming problems", *Management Science*, **8** (4), 419–41, July 1962.

This represents an article with the stated title by C. G. Boot which appeared on pp. 419–41 in issue number 4 of Volume 8, dated July 1962, of the periodical *Management Science*.

Odd as it might at first seem, information on a particular *subject* can often be traced by consulting *author* indexes. The author of a useful paper which has already been traced may well have written others of equal or even greater value. Furthermore, it is as a rule easier to check author indexes than it is to check those arranged by subject.

When a reference has been located in the volume-index of the periodical in question, then consulting the full article is a simple matter. It is merely a question of turning to the appropriate pages in the (bound) volume. If, on the other hand, the reference has been noted in an abstracting or indexing publication, or found in a bibliography, then it will be necessary first to locate a copy of the original document. This is usually a case of discovering a library which possesses a file of the periodical concerned.

Directories of the combined periodical holdings of a large number of libraries have been compiled to facilitate the location of individual titles. The *Union List of Serials in Libraries of the United States and Canada*, edited by Winifred Gregory, has an alphabetical arrangement of titles and indicates the holdings of each library represented by means of symbols, to which there is a key. The second edition was published in 1943 by Wilson, New York. There are also two supplements which bring the period covered up to December, 1949, and these volumes are updated by a Library of Congress publication called *New Serial Titles*. A similar service is provided by the *British Union Catalogue of Periodicals*, 4 volumes, Butterworths, London, 1955–8. Its subtitle is "A record of the periodicals of the world from the seventeenth century to the present day, in British libraries". A supplement, bringing the coverage up to 1960, was published in 1962.

Other directories, such as the *World List of Scientific Periodicals*, are listed in Brummel and Egger's *Guide to Union Catalogues and International Loan Centres*, mentioned on p. 17.

A great deal of time spent in literature searching can be saved by the systematic perusal of current periodicals coupled with the creation and maintenance of one's own information index. It is both frustrating and wasteful of time to have to search for the particulars of a paper which one has already read but omitted to index.

Abstracting and Indexing Publications

The days have long since passed when a mathematician could reasonably hope to be acquainted with most if not all of the important literature of his subject. With the increasing applications of mathematics has come more intensive specialization by its practitioners. Probably Poincaré was the last of the mathematicians who could fairly be considered to be abreast of the advancing front along its whole length. Greater specialization in one field has led to further remoteness from the others. As new sub-fields are developed to cope with the requirements of our techno-logical age so the amount of literature grows in total volume. Each topic of specialization acquires its own bibliographical resources which in turn divide and grow in amoeba-like fashion as the process develops. The problem confronting the mathe-matician is twofold: how to keep track of the progress made both in his own and other branches of the subject, and how to trace back through the welter of literature and pinpoint the particular facts or theories which he may need from time to time in this work.

This situation is prevalent throughout the twin fields of science and technology. In fact the mathematician is looked to for a solu-tion to the problem of what is now termed *information storage and retrieval*. Computers are being specially designed to digest biblio-graphical information and to reproduce it on demand according to individual requirements and specifications. It will be some years yet, however, before the computer completely replaces the manual methods now in use. The printed abstracting publication has not yet in fact reached its full development.

Abstracts of literature are generally considered to fall into two classes: indicative and informative. The former simply sets out the bibliographical details of the original (author, title, publisher, date, etc.) and gives a broad indication of the subject matter. The informative abstract, on the other hand, includes the essential "meat" of the original, reproducing factual data, the steps of experiments, and the main lines of theories. It can sometimes obviate the need to consult the original, and is especially valuable in the case of material which is not readily accessible or is perhaps written in a foreign language. As a rule, however, it is unwise to rely solely on abstracts as they are naturally subject to bias on the part of the abstractor. When an abstract includes an evaluation of the original it is properly called a *review*.

Abstracting publications usually appear in the form of periodicals, though some are issued on index cards which can be selectively filed to suit the individual requirements of the subscriber. Their major use is in documenting periodical articles. The number of abstracting services has grown with the number of new periodicals. The more periodicals there are the more abstracting services there will be; and the more services there are the more those periodicals will be used.

Abstracts are produced by different kinds of organizations. In *A Guide to U.S. Indexing and Abstracting Services in Science and Technology* (National Federation of Science Abstracting and Indexing Services) nearly 500 publications are listed, of which some 50 per cent are issued by commercial bodies, about 40 per cent by societies and institutions, and the remaining 10 per cent by the U.S. Government.

The literature of mathematics from the second half of the 19th Century to the present day is covered by the following publications:

> *Bulletin de Bibliographie, d'Histoire et de Biographie Mathématiques* (8 volumes), 1855–62.
> *Bullettino di Bibliografia e di Storia delle Scienze Matematiche e Fisiche*, 1868–87.
> *Jahrbuch über die Fortschritte der Mathematik*, 1868–1934.
> *Revue Semestrielle des Publications Mathématiques*, 1893–1934.
> *Zentralblatt für Mathematik*, 1931–.
> *Mathematical Reviews*, 1940–.
> *Referativyni Zhurnal—Matematika*, 1953–.

A useful guide to earlier work is the Royal Society of London's *Catalogue of Scientific Papers, 1800–1900* published by Cambridge University Press. Volume 1 of its subject index, 1908, covers pure mathematics.

Mathematical Reviews, published by the American Mathematical Society, is undoubtedly the most valuable current English-language publication in the field. It is issued monthly. Its subject coverage is extremely wide, and it is international in scope. Some entries are reprinted from other services whose titles should be noted:

Applied Mechanics Reviews,
Computing Reviews,
Mathematics of Computation,
Operations Research,
Referativyni Zhurnal,
Science Abstracts,
Zentralblatt für Mathematik.

Each of these is an important abstracting publication in its own right. In addition to the American Mathematical Society, over a dozen other national mathematically oriented societies sponsor the publication which also has financial aid from the National Science Foundation. This is such an important tool for the mathematician that the majority of the headings used in the arrangement of the reviews are reproduced here:

General, history and biography; logic and foundations. set theory, combinatorial analysis, order, lattices, general mathematical systems; theory of numbers, fields and polynomials, abstract algebraic geometry, linear algebra, associative rings and algebras, non-associative algebra, homological algebra; group theory and generalizations, topological groups and Lie theory; functions of real variables, measure and integration, functions of a complex variable, potential theory, several complex variables, special functions, ordinary differential equations, partial differential equations, finite differences, and functional equations; sequences, series, summability, approximations and expansions, Fourier analysis, integral transforms, operational calculus, integral equations, functional analysis, calculus of variations; geometry, convex sets and geometric inequalities, differential geometry, general topology; algebraic topology, topology and geometry of differential manifolds; probability, statistics, numerical methods, computing machines.

A separate index issue is published for each six-monthly volume. Some idea of the exhaustiveness of this work is given by the fact that in the first six months of 1962, 4245 reviews were published.

The *Bulletin Signalétique* of the Centre de Documentation du Centre National de la Recherche Scientifique in Paris is published in 22 sections, of which the first is entitled *Mathématiques*. It appears monthly. The abstracts are arranged in a classified sequence, and an author index is included in each issue.

The purely mathematical abstracting publications may be supplemented, especially for subjects of marginal interest, by compilations of a more general character. *Engineering Index*, for example, contains some 30,000 brief abstracts every year selected from about 1400 different periodicals. After 1957, the *Industrial Arts Index* (Wilson, New York) split into the *Applied Science and Technology Index* and *Business Periodicals Index*. Both index articles in a large number of periodicals, the titles covered being in fact decided by a vote of subscribers. 1962 saw the inception of *British Technology Index*, published monthly by the Library Association, London.

Cover-to-Cover Translations

Over the past few years several societies and commercial organizations, usually with strong government backing, have undertaken the complete translation of important foreign periodicals on a current basis. Although annual subscription rates tend to be rather high in some cases, one of the objects of such publications is to reduce the total amount of money spent on individually commissioned translations. But it is more than a simple cost-sharing process. Complete translations of periodicals make personal selection of articles unnecessary. If items for translation were selected according to the probable demand for them it is quite likely that material which at present would appear to have little use would remain untranslated, and therefore possibly overlooked by later workers to whom it might be of value. A relevant statement by Aslib is quoted on p. 133.

Most cover-to-cover translations concern Russian-language periodicals and are therefore dealt with in Appendix I. The American Mathematical Society has however recently commenced publications of *Chinese Mathematics—Acta*, which is a complete translation of *Acta Mathematica Sinica*, the official journal of the Institute of Mathematics of the Academy of Sciences, Peking. It contains papers on research in pure and applied mathematics. The first translated volume corresponds with Volume 10, 1960, of the original.

Exercises

1. List the uses to which an abstracting publication can be put. Assume that both the current issue and a complete back file are available.
2. Why is it important for a mathematician to peruse a number of periodicals as a regular part of his work?
3. What is meant by the term *cover-to-cover translations*, and why are they considered necessary?
4. If you were given a grant to cover subscriptions to three mathematical periodicals, how would you choose the titles?

CHAPTER 5

Mathematical Societies

510.6: Mathematical societies

ANY differences which may exist between societies, associations, and institutions need not be considered here. All are treated together as organizations comprising memberships of persons interested or actively engaged in and devoted to the furtherance of a particular subject. For convenience, they will be referred to as societies.

An outline of the history and rise of mathematical societies will be found in most large general encyclopedias. As the rate of scientific and technical progress has increased and research has become more specialized, so the related societies have grown in number and become more restricted in their subject scope. Early societies, such as the Royal Society of London, covered a wide range of scientific disciplines. Some of those which have survived have divided their activities into sections in order to enable them to deal more efficiently with each individual subject. Newer societies are almost invariably devoted to quite narrow subject fields.

Functions

All societies publish a statement of their aims and purpose, and indicate the nature of their membership. Sometimes, conditions for membership are very stringent and demand high academic attainment. Membership, especially in the case of those bodies which bestow qualifications based on examinations or other criteria, may be at various levels according to the individual member's state of advancement. This arrangement is clearly shown in the Institute of Actuaries, London, where provision is made for membership in the grades of student, associate, and fellow. Generally speaking, specialized societies are anxious to encourage students, and are pleased to answer enquiries, within their capabilities, from any bona fide enquirer.

An analysis of the publications of the major mathematical societies reveals the following as their main areas of activity:

Meetings—local, national, international,

Publications—periodicals (including proceedings and transactions), books, monographs, etc.,

Library and information services,

Educational programmes,

Vocational guidance services,

Co-operation with other organized bodies.

Societies are often called upon to advise on problems concerning mathematics which arise in a wide variety of situations. Their influence can be quite considerable, for example, in determining the course of official enquiries into matters affecting mathematical education. Referring to mathematics in industry, J. T. Combridge, President of the Mathematical Association, London, recently said in a letter to fellow-members: "There are large companies who have the vision to realise that they need a body like our Association to advise them . . .". It will be found that true scholars who dedicate themselves to the advancement of their subject are invariably willing to give others the benefit of their knowledge and experience.

Advantages of Membership

No society can function without members, and the advantages of membership should therefore be seen as the product of a co-operative effort. This does not simply mean the payment of dues, but the active support of as many of the society's activities as possible. The society, by virtue of the nature of its constitution, may attract other sources of income such as government grants, contributions from industry, and institutional membership.

Reciprocal arrangements may be made with societies in other countries whereby certain privileges, such as reduced prices of publications, may be extended in both directions. The American Mathematical Society, for instance, has entered into reciprocity agreements with nearly thirty overseas mathematical organizations.

It cannot be too strongly emphasized that an important complement to bibliographical sources of information is the fund of knowledge possessed by subject specialists. The advantage of society membership in this direction is immeasurable, for it provides a ready means of contacting such specialists and becoming aware of their respective fields of interest. Most societies issue membership directories, which are supplemented by biographical notes in their periodical publications.

International Organizations

The Union of International Associations (Palais d'Egmont, Brussels 1, Belgium) publishes a directory entitled *Yearbook of International Organizations*. Improved communications have led to an increasing number of international meetings in a wide range of subjects. Much valuable work is done, but due to inadequate documentation of resultant publications this has remained a largely untapped source of information. In an endeavour to improve the situation the Union commenced publication, in January 1961, of *Bibliographie Courante des Documents, Comptes Rendus et Actes des Réunions Internationales*. Another of its publications is *International Congress Calendar*. The 1962 edition gives details of scheduled international meetings in all fields for the years 1962–7.

The *World List of Future International Meetings* is published by the Library of Congress, Washington, D.C.

International mathematical conferences are held every four years. Two valuable prizes are awarded at each conference for the best work done by a mathematician under 40 years of age. In 1962, the International Congress of Mathematicians was held in Stockholm, with the Swedish National Committee for Mathematicians and the Swedish Mathematical Society acting as hosts. The first International Congress was held in 1897. The principal international mathematical body is the International Mathematical Union, Eidgenössische Technische Hochschule, Zurich, Switzerland. The object of the 37 member countries is to foster international co-operation in mathematics. The Union assists the International Congress of Mathematicians and encourages other activities which are intended to stimulate the development of mathematics.

National Organizations

National mathematical organizations cover the three major areas of research, education, and applications. In some countries all these areas are gathered under the wing of one society, in others different societies cater for each separately. They normally issue, free of charge, a brochure detailing their aims, activities and conditions of membership. Other brochures will deal with such topics as careers, education, and publications. Regular publishing activities vary considerably between societies, but most national bodies issue, as a minimum, a journal in which their transactions and proceedings appear. The contributions to such periodicals and the content of any other scientific literature published by national societies can usually be relied upon to be authoritative. When information is required on any branch or aspect of mathematics in a foreign country, the national society will be the first source to approach.

The Moscow Mathematical Society (Moskovskoe Matematicheskoe Obshchestvo) is one of the oldest of the major national mathematical societies. Founded in 1864, it publishes *Matematicheskiĭ Sbornik* and its transactions, *Trudy*. In the following year came the London Mathematical Society, and in 1894 the American Mathematical Society which actually began in 1888 as the New York Mathematical Society.

The following are all national bodies and are arranged alphabetically by country:

Argentina: Union Matemática Argentina
                Casilla de Correo 3588,
                Buenos Aires.

Australia: Australian Mathematical Society
                Department of Mathematical Statistics,
                University of Sydney.

Austria: Österreichische Mathematische Gesellschaft
                Karlsplatz 13,
                Vienna IV.

Belgium: Société Mathématique de Belgique
                317 Avenue Ch. Woeste,
                Brussels 9.

Canada: Canadian Mathematical Congress
                Ecole Polytechnique,
                Montreal, Quebec.

China: Chinese Mathematical Society
                P.O. Box 143,
                Taipei, Taiwan.

Denmark: Dansk Matematisk Forening
                Blegdamsvej 15,
                Copenhagen.

Finland: Suomen Matemaattinen Yhdistys
                University of Helsinki,
                Hallituskatu 11–13,
                Helsinki.

France: Comité National Français de Mathématiciens
                11 rue Pierre Curie,
                Paris 5e.
                Fédération Française des Sociétés de Mathématiques—
                address as above.
                Société Mathméatique de France—
                address as above.

Germany: Berliner Mathematische Gesellschaft e.V.
                Berlin—Charlottenburg 2,
                Hardenbergstrasse 34.
                Deutsche Mathematikervereinigung e.V.
                Tübingen,
                Math. Institut der Universität.

Italy: Unione Matematica Italiana
                Istituto Matematico dell'Universita,
                Largo Trombetti, 4, Bologna.

Japan: Mathematical Society of Japan
                Faculty of Science,
                University of Tokyo.

Mexico:	Sociedad Matemática Mexicana
	Tacuba No. 5,
	Mexico 1, D.F.
Netherlands:	Wiskundig Genootschap
	Amsterdam C.
Norway:	Norsk Matematisk Forening
	Matematisk Institutt,
	Universitetet, Blindern,
	Oslo.
Peru:	Sociedad Matemática Peruana
	Apartado 4963,
	Lima.
Poland:	Polskie Towarzystwo Matematyczne
	ul. Sniadeckich 8,
	Warsaw.
Scotland:	Edinburgh Mathematical Society
	Mathematical Institute,
	16 Chambers Street,
	Edinburgh, 1.
	Glasgow Mathematical Association
	Department of Mathematics,
	The University,
	Glasgow 2.
Spain:	Real Sociedad Matemática Española
	Serrano 123,
	Madrid.
Sweden:	Svenska Matematikersamfundet
	K.T.H.,
	Stockholm 70.
Switzerland:	Schweizerische Mathematische Gesellschaft
	E.T.H.,
	Zurich.
	Stiftung zur Förderung der Mathematischen Wissenschaften in der Schweiz
	Genferstrasse 2,
	Zurich.
U.S.S.R.:	Academy of Sciences of the U.S.S.R.
	Department of Physics and Mathematics,
	Pyzhevsky per. 3,
	Moscow.
U.K.:	London Mathematical Society
	Burlington House,
	London, W.1.
	Mathematical Association
	Gordon House, 29 Gordon Square,
	London, W.C.1.

E

U.S.A.:   American Mathematical Society, and
Association for Symbolic Logic
190 Hope Street,
Providence 6, Rhode Island.
Industrial Mathematics Society
100 Farnsworth Street,
Detroit 2, Michigan.
Mathematical Association of America
University of Buffalo,
Buffalo 14, New York.
Society for Industrial and Applied Mathematics
P.O. Box 7541,
Philadelphia 1, Pennsylvania.

Tracing Societies

Tracing societies is facilitated by the existence of directories which specialize in this kind of information. Difficulties may sometimes be experienced where a society has no fixed headquarters, and the only address that can be located is that of an elected secretary or president whose term of office may have expired. When directory information is inadequate, recourse has to be made to the national society. This may also be the case with newly created societies of which the particulars have not yet appeared in directories. Guides to societies vary in the amount of detail they provide. Some go as far as indicating the total membership and the names of the principal officers. Those which list the societies' publications are useful. Others, however, content themselves with names and addresses. Usually the entries are grouped under broad headings with indexes giving a complete alphabetical list of the societies and a subject guide to their activities. Directories which are frequently revised, either as completely new editions or by the publication of supplements, are obviously to be preferred, but it should be remembered that any directory is out of date even on the first day of publication. When writing to a society it is therefore usually best to address either the society or the secretary impersonally rather than an individual officer.

Some useful directories are as follows:

Europa Year Book, 2 volumes. Europa Publications, London, Annual.

Volume 1 covers Europe, and Volume 2 Africa, the Americas, Asia and Australasia. Under each country is a directory section, where information useful to the mathematician includes insurance companies, research institutes, libraries and universities.

Encyclopedia of Associations. Gale, Detroit, 3rd edition, 1961.

Volume 1 covers national organizations of the United States.

Scientific and Learned Societies of Great Britain, a Handbook compiled from Official Sources. London, Allen & Unwin. 60th edition, 1962.

Part I covers government and public bodies conducting scientific research in Great Britain. Part II is a list of scientific and learned societies. It is divided into 16 sections of which Section 1: *General Science*, and Section 11: *Mathematical and Physical* are of most interest to the mathematician.

Scientific and Technical Societies of the United States and Canada, Washington, National Academy of Sciences—National Research Council. 7th edition, 1961.

The societies are arranged alphabetically, and there is an analytical index to them. Particulars include history, purpose, membership, meetings, library, research funds and publications.

The names, addresses, and activities of societies in related fields are given in later chapters together with other information in the respective subjects.

Society Library and Information Services

Many societies maintain libraries for the use of members. Those belonging to old-established bodies may be considerable in size and very comprehensive. They form unique special collections in their subject fields, and their catalogues constitute excellent subject bibliographies. Most societies are prepared to allow enquirers who are not members to use their libraries. This function does indeed make an important contribution to the increasing volume of inter-library loans in which all types of libraries participate.

Those societies which publish journals having book review sections are in a very strong position with regard to book accessions, since they receive many review copies of new publications. Societies also frequently enter into exchange agreements with other learned bodies. Usually periodical publications are the subject of such agreements. In January 1962 a U.S.–U.S.S.R.

exchange agreement came into effect, under which the American Mathematical Society and the Library of the Academy of Sciences of the U.S.S.R. undertook to ensure the supply of each other's mathematical journals and book series. Initially, some 1000 subscription exchanges were involved. UNESCO has published a second edition of its *Handbook on the International Exchange of Publications* which lists libraries with publications available for exchange. The list is updated by a regular section in the *UNESCO Bulletin for Libraries*. Among recent listings is the Society of Mathematical and Physical Sciences of the Rumanian People's Republic, offering its *Buletinul Matematic* and other publications in exchange for publications on the same subject. Not all the participants, of course, are societies. Another example is the Mathematical Institute of the Serbian Academy of Sciences and Arts which offers its periodicals in exchange for other publications on mathematics. Donations of personal collections from members are yet another valuable source of stock acquisition for society libraries.

The library of the Mathematical Association, London, may be described by way of example. It is housed in the Library building of the University of Leicester. Members of the Association may borrow books and journals for periods not exceeding three months at a time, on payment of the cost of postage. A full printed catalogue of the collection is unfortunately not available, but four printed lists of *Books and Pamphlets in the Library of the Mathematical Association* together cover accessions up to the end of 1935.

Another example is that of the Royal Statistical Society, whose library comprises some 80,000 volumes. About 350 different periodicals are currently received. The library covers the entire range of statistics, including the employment of statistical methods in industry.

Other British examples will be found in the *Aslib Directory*, Aslib, London, 1957. American society libraries are included in the *American Library Directory* and *Subject Collections*, both described in Chapter 2, and in *Scientific and Technical Societies of the United States and Canada* mentioned above. Most industrial countries of the world have active special libraries (society, industry, government, etc.) and these are the subject of directories. Australian special libraries, for example, are covered by *Directory of Special Libraries in Australia*, Library Association of Australia, Sydney, 2nd edition, 1961. Up-to-date information on the society libraries of other individual countries may be sought from the respective national library associations.

Locating Society Publications

Locating society publications in catalogues and bibliographies can present certain problems of which it is important to be aware. It is not the present purpose to deal with cataloguing procedures, but their implications can often hamper the layman. The difficulty revolves around the form of author heading which is used. A society publication may be produced either by the society itself, or by an individual or individuals on behalf of the society. In the latter case, although recognized cataloguing codes give specific rulings on the point, some compilers of bibliographies adopt the method of using the society's name as the author entry while others use the name of the individual as the entry word. Confusion also arises in the alternative treatment given to the form of entry used in listing the periodical publications of a society. The most widely adopted rule here is that when the title of the periodical is distinctive and does not include the name of the society, then the entry simply consists of the title of the publication itself. An example is the *Journal of Symbolic Logic*, which is the official organ of the Association for Symbolic Logic. Where, however, the name of the society does figure in the title, this comes first. Thus the *Proceedings of the American Mathematical Society* would be listed as *American Mathematical Society, Proceedings*. One of the advantages of this method is that the exact title of a particular society's publication need not necessarily be known in order to locate it. One would otherwise have to proceed by trial and error using the numerous possible alternatives: Journal, Proceedings, Transactions, Annals, and so forth. This is even more hazardous when a foreign title is in question. Nevertheless, practice is by no means uniform in the major bibliographical works. The *World List of Scientific Periodicals* and the *British Union Catalogue of Periodicals* have adopted different systems. When using such tools it is essential to read the editor's instructions. These will also deal with the way in which title changes have been catered for. Care must always be taken to try all possible alternatives before abandoning a search.

Exercises

1. Write notes on any four activities of mathematical societies, quoting actual examples where possible.
2. Discuss the advantages and responsibilities attaching to membership of a professional society.
3. In what respects are the library resources of a mathematical society complementary to those of a large public library?
4. Name some of the difficulties of locating society publications in catalogues and bibliographies.

CHAPTER 6

Mathematical Education

510.7: Mathematics, study and teaching

EVERY professional mathematician has a responsibility towards the advancement of his subject and the spread of mathematical knowledge. He is concerned with one aspect of education or another throughout his career. Many qualified men and women who are primarily occupied in non-educational work undertake part-time lecturing. Others contribute articles to periodicals, write books, or simply take part in discussions at society meetings. It is essential to the advancement of science that the fruits of the latest research should be digested into the main body of mathematical education.

The principal responsibility in this process clearly lies with those actively engaged in teaching, from the university professor down to the primary school teacher. They must be aware of the most efficient pedagogical techniques and at the same time absorb as much as they can of the increasing bulk of mathematical literature. If they are to succeed they need to be thoroughly conversant with the appropriate sources of information in both fields.

Purely educational matters lie outside the scope of the present work, but attention should be drawn to the fact that in the Dewey Decimal Classification the literature of education is catered for in the 370 class. Often, in classifying educational literature, its level is a deciding factor. This may be exemplified by reference to textbooks. In the 16th edition of Dewey, provision is made for textbooks on a specific subject *at elementary school level* at 372.3–372.8. Textbooks on a specific subject *at higher levels*, however, are placed with that subject, for example, algebra 512.

61

It is for this reason that textbooks are here dealt with in Chapter 9. Secondly, it should be observed that many of the works referred to in the following sections would, in a general library, normally be classified under education.

Mathematics Courses and Financial Aid

Grants are made to students by a number of different bodies, among which universities, colleges and governments figure most prominently. The American Mathematical Society publishes an annual survey of *Assistantships and Fellowships in Mathematics* as a special issue of its *Notices*. It lists institutions in the United States and Canada offering degree courses in mathematics and closely allied subjects with details of assistantships and fellowships. There is a complete alphabetical index of departments of mathematics, statistics and applied mathematics. The *New American Guide to Colleges*, by Gene R. Hawes (Columbia University Press, 1962) gives information on 2000 U.S. colleges and universities, including admission policy and financial aid. A useful feature is its "College Discovery Index".

More and more students are undertaking studies in overseas countries. Officially sponsored programmes aim to facilitate this, particularly in the field of student exchanges. The growth has been substantial enough to warrant the publication of a new journal entitled *Overseas: the Magazine of Educational Exchange* (1961). UNESCO has done much to foster foreign study. Every year it publishes a directory entitled *Study Abroad*. The 1962 edition contains information on over 115,000 individual opportunities for international study and travel offered by 1674 awarding agencies throughout the world. The proportional distribution of awards is given as about 45 per cent for post-graduate study and research, some 30 per cent for study towards university degrees, and the remaining 25 per cent covering other forms of educational activity. The number of awards in mathematics is high. A chapter also gives particulars of nearly 300 organizations offering advisory services and practical help to persons wishing to study abroad.

Fulbright Travel Grants are well known, and have assisted a great many students.

Notes on Grants Awarded by the D.S.I.R. (Department of Scientific and Industrial Research) *to Research Workers and Students* is published annually by H.M. Stationery Office, London.

Every university, college or other educational institution issues its own prospectus, often called a *calendar*, in which particulars of courses offered, admission requirements, and financial assistance are among the information given. Still greater detail is available in the publications of individual departments.

There are several comprehensive guides to universities and colleges covering different areas. Among these are:

> *International Handbook of Universities*, International Association of Universities, Paris, 1959.
>
> *American Universities and Colleges*, American Council on Education, Washington, D.C., 8th edition, 1960. Described as "a catalog of 1,058 college catalogs".
>
> *Commonwealth Universities Yearbook*, Association of Universities of the British Commonwealth, London. Among several useful appendices are two covering university admission requirements, and the Commonwealth Scholarship and Fellowship Plan.

Need for Teachers

In countries such as the United States and Great Britain, out of the total number of qualified mathematicians approximately one half find employment in academic institutions. In Britain, the majority of university trained mathematicians are employed in teaching of one kind or another. The actual number is over 10,000. To this figure must be added many more who have not passed through university. The new Diploma in Mathematics awarded by the Mathematical Association is primarily intended for non-graduate teachers. Recent figures for the United States show a tendency for younger mathematicians to seek employment outside teaching. In fact, while the proportion of all mathematicians in educational institutions is 42 per cent, the figure for those under 40 years of age is only 36 per cent.

There is an undoubted shortage of mathematics teachers, and the position is being aggravated by the increasing opportunities for less qualified mathematicians in business and industry. These two fields combined are showing only a slightly lower intake of qualified people than teaching. The situation has now arisen when mathematics enrolments at all levels of education are increasing, while the proportion of qualified mathematicians entering the teaching profession is diminishing. More mathematics teachers are required not only to train potential mathematicians but also to provide adequate mathematical training for the other scientists and technologists whose services are in such great demand, especially physicists, chemists and engineers.

Teacher Training

Teacher qualifications vary according to the level at which the subject is to be taught. They also differ in certain respects from one country to another. An excellent survey of European and American practices is given in the O.E.E.C. publication mentioned on p. 67. Chapter 7 of that work, entitled "The Training of Teachers", deals with the training of elementary and secondary school teachers, pedagogical requirements, present supply and future demand of teachers, and maintaining professional competence.

British conditions are covered by the following three publications which are all available free of charge from the Ministry of Education, Curzon Street, London, W.1:

Becoming a Teacher
A Career in Education for University Graduates
Courses of Professional Training for Teachers and Intending Teachers in Technical Colleges and Similar Establishments of Further Education

Another useful publication, which deals with conditions in Scotland, is entitled *Prospects in Teaching* and is available upon application to the Scottish Education Department, St. Andrew's House, Edinburgh 1.

For America there is the *Occupational Outlook Handbook* prepared by the U.S. Department of Labor. Its compilation involved a survey of 650 occupations, but separate reprints of individual sections may be obtained. That dealing with teaching costs 15 cents. *Mathematics Teaching as a Career* is available free from the National Council of Teachers of Mathematics.

The American Mathematical Society's annual survey of *Assistantships and Fellowships in Mathematics* gives details of master's and doctoral degrees designed for the teaching of mathematics. These are offered at nearly 200 institutions and branches of institutions. A new degree, Master of Arts in Teaching (M.A.T.), is offered at some two to three dozen institutions. Almost three-quarters of the institutions make it possible to earn a degree through summer study alone.

To assist in the adoption of new changes in curricula, the U.S. National Science Foundation operates a programme of institutes for high school mathematics teachers. A relevant publication is Andrée, R. V., *The Road for Modern Mathematics*. N.S.F. Summer Institute for Teachers of Secondary School Mathematics, Oklahoma Agricultural and Mechanical College, Stillwater, 1955.

The Journal of Teacher Training, published by the National Education Association, Washington, 6, D.C., may be used for keeping in touch with matters of current interest.

Mathematical Education Periodicals

Periodicals concerned with mathematics teaching should be regularly perused in order to keep abreast of the latest techniques. Full publication details are given here of some English-language journals.

> *Arithmetic Teacher* and *Mathematics Teacher*
> National Council of Teachers of Mathematics,
> 1201 Sixteenth Street, N.W.,
> Washington 6, D.C., U.S.A.
> *Australian Mathematics Teacher*
> The Mathematical Association, N.S.W. Branch,
> c/o Sydney Teachers' College, University Grounds,
> Newton, N.S.W., Australia

Mathematical Gazette (Journal of the Mathematical Association)
G. Bell and Sons,
Portugal Street, Kingsway,
London, W.C.2, England

Mathematics Teaching, published for the Association for Teaching Aids in Mathematics by Ian Harris, 122 North Road, Dartford, Kent, England.

School Science and Mathematics
Central Association of Science and Mathematics Teachers, Inc.,
P.O. Box 108, Bluffton, Ohio, U.S.A.

There are also several important foreign-language periodicals from which the following have been selected:

Belgium:	*Mathematica et Pedagogia*
	Mathésis
France:	*Association des Professeurs de Mathématiques de l'Enseignement Public, Bulletin*
	L'Enseignement des Sciences
Germany:	*Mathematische und Naturwissenschaftliche Unterricht*
	Mathematische–Physicalische Semester Berichte
	Archimedes
Italy:	*Archimede*
	Scuola e Didattica
Netherlands:	*Euclides*
Portugal:	*Gazeta de Mathemática*
Scandinavia:	*Nordisk Matematisk Tidsskrift*
Switzerland:	*L'Enseignement Mathématique*
	Elemente der Mathematik

Further details both of these and of other relevant titles may be found in two UNESCO publications:

World Survey of Education, Volume 3: *Secondary Education*, Paris, 1961.

Educational Studies and Documents, No. 23: *An International list of educational periodicals*.

Periodicals written from the viewpoint of the student are also published, such as *Student Mathematics Journal*, National Council of Teachers of Mathematics, Washington, D.C.

Curricular Changes

Mathematical education has remained rather static for a good many years. It is in fact said that the mathematics taught in schools at the present day is virtually the same as that taught two centuries ago. The need for a drastic revision of techniques is, however, being more widely recognized. New curricula are being put forward, based on critical surveys of current practices.

Evidence of the introduction of revised programmes in the United States is reflected, for example, in the examinations of the College Entrance Examination Board and the Educational Testing Service. The bodies mainly responsible for the changes are the School Mathematics Study Group, the University of Illinois Committee on School Mathematics, the Ball State College Project and the Commission on Mathematics.

In Great Britain, a significant event was a conference convened at Southampton in 1961. The findings of the delegates were reported in:

> Southampton Mathematical Conference, 1961, *On Teaching Mathematics: a Report on Some Present-Day Problems in the Teaching of Mathematics,* Pergamon Press, Oxford, 1961.

On the international level, a conference was organized by the Office for Scientific and Technical Personnel of the Organization for European Economic Co-operation. An edited version of the papers, discussions, and recommendations may be found in:

> O.E.E.C. *New Thinking in School Mathematics.*
> The Organization, Paris, 1961.

Teaching Methods

If the method of teaching mathematics has not changed in 200 years this is not to say that the science itself has stagnated. On the contrary, progress in the last 50 years alone has outstripped the entire product of several centuries. Unfortunately, the new advances have not been incorporated evenly throughout the whole range of educational levels. At the higher levels of university education account is being taken of recent research, and increasing attention is also being paid to adjusting undergraduate courses accordingly. There remains, however, a dislocation between secondary school and university instruction. This is the centre of the problem, which is being tackled quite vigorously in certain countries. What is required is an adjustment of the type of mathematical thinking among students. Such adjustment requires new pedagogical methods. Mathematics has developed a new language. It is commonly felt that there is a need to teach modern symbolism in secondary schools as early as possible. Teachers themselves need to be fully conversant with the concepts of what is commonly referred to as *the new mathematics*. Useful publications on this subject are available, among them the following:

ADLER, IRVING, *The New Mathematics*, Dobson, London, 1959.

BANKS, JOHN HOUSTON, *Elements of Mathematics*, Allyn and Bacon, Boston, 2nd edition, 1961.

COURANT, RICHARD and ROBBINS, HERBERT, *What is Mathematics? An Elementary Approach to Ideas and Methods*, Oxford University Press, New York, 1941.

Contains 4 pages of "Suggestions for further reading".

National Council of Teachers of Mathematics. *Insight into Modern Mathematics*, The Council, Washington, D.C., 1957.

SAWYER, WILLIAM WARWICK, *Prelude to Mathematics*, Penguin Books, 1955.

This gives the ideas behind mathematics, and deals especially with matrices and determinants.

STABLER, E. R., *Introduction to Mathematical Thought*. Addison-Wesley, Cambridge, Mass., 1953.

There are many useful books on teaching techniques. A British series edited by the Mathematical Association and published by Bell includes *Teaching Algebra in 6th Forms*, and *Teaching Geometry in Schools*. In 1959 the same publisher issued *Mathematics in Secondary Modern Schools: a report prepared for the Mathematical Association for consideration by all concerned in the teaching of mathematics in secondary schools*. *The Teaching of Mathematics*, Cambridge University Press, 1957, was produced by the Incorporated Association of Assistant Masters in Secondary Schools. Pamphlet 38 of the Board of Education, 1958, is entitled *Teaching Mathematics in Secondary Schools*.

A bibliography will be found in G. Mialaret's "Teaching of

Mathematics", *Education Abstracts* (UNESCO), **11**, (9), Nov. 1959. Its 79 annotated entries are arranged under country headings. Articles in periodicals have generally been excluded, and apart from works of outstanding importance the bibliography lists no works prior to 1950 and is brought up to about Easter 1959.

Audio-visual methods and learning machines are rather too specialized for the present work.

It should be remembered that professional associations are always prepared to offer assistance in locating information, and to recommend appropriate literature. Several addresses are given in the section on mathematical education periodicals, and No. 38 of UNESCO's series of *Educational Studies and Documents* is an *International Directory of Education Associations*.

Exercises

1. This chapter has many links with the four preceding chapters. Write brief notes on some of these, and quote the titles of publications given in the previous chapters to illustrate your answer.
2. What is the main reason for the revised mathematical curricula which are now being introduced?
3. What steps can the mathematics teacher take to keep abreast of the new developments in teaching methods in his subject?

CHAPTER 7

Computers and Mathematical Tables

510.78: Computation instruments and machines
510.83: Mathematical tables

COMPUTER literature has received fresh treatment in the 16th edition of the Dewey Decimal Classification. 510.78 is devoted to Computation Instruments and Machines, which are divided into Analogue Machines (510.782) and Digital Machines (510.783). Their actual manufacture, however, is catered for at 681.141–681.144. The standard 15th edition had allocated 510.81 to Computation Instruments and Machines, but made the proviso that where interest lay chiefly with electronics, the appropriate class was 621.34. Neither 510.78 nor 510.81 had been used for computer literature in the 14th edition, but 681.14 accommodated Calculating, Counting and Registering Machines.

Whilst all this appears rather confusing, it is in fact a good example of the type of problem which besets the classifier who has to cope with the literature of a subject whose development is outpacing that of his classification schedules. As far as the library *user* is concerned, the catalogue will indicate how the various aspects of computers and computation have been treated.

Computers

Orientation

Electronic computers have brought about an industrial revolution. In many industries they have completely changed the character of research, being applied to the solution of complex problems which might not otherwise have seemed capable of solution. This is the case, for example, in today's vitally important research in aerodynamics, where factors governing the design of the latest supersonic aircraft and space vehicles are calculated with the aid of computers. Indeed, new applications are being devised to meet the requirements of an ever widening range of scientific and technological endeavour. The design, development and application of computers have stimulated a vast amount of basic research in mathematics, demanding a constant supply of fully qualified mathematicians.

In its turn, this research has generated a steady flow of new literature of all forms: handbooks, textbooks, periodicals, technical reports, tables, conference proceedings, standards and patents. Every aspect of the subject has received its share of attention, from mathematical theory to the most detailed refinements of the applications of individual machines to individual problems.

The difficulty for the student or layman is to know where to start. A distinction must first be drawn between the literature dealing with the manufacture of computers and that covering theoretical design and ultimate applications. Manufacture is largely the province of the instrument engineer, whilst the other aspects are of varying degrees of interest to the mathematician. The distinction is reflected in the schedules of the Dewey Decimal Classification, though in general bibliographies this is not always the case. In *Electronics: a Bibliographical Guide*, by C. K. Moore and K. J. Spencer (Macdonald, London, 1961), the entries are arranged according to the Universal Decimal Classification, but no specific provision is made for mathematics. All items relating to computers are therefore grouped under the 681.14 heading.

Standard reference works on computers must necessarily cover every facet of the subject. This is true of Harry H. Husky and Granino A. Korn's *Computer Handbook* published by McGraw-Hill in 1962. The essential mathematical theories are included, and extensive lists of references are appended to each of the sections into which the work is divided. Another recent handbook is

F

Francis J. Murray's *Mathematical Machines*, Columbia University Press, 1961. It is in two volumes, the first entitled *Digital Computers*, and the second *Analog Devices*. Again, references follow each chapter.

An unusual book is *The Arithmetic of Computers* by Norman A. Crowder, English Universities Press, London, 3rd edition, 1962 (first published in the U.S.A. by Doubleday). It is issued under the trademark *TutorText*, and the presentation is intended to simulate a conversation between teacher and student. The pages are not read consecutively. The reader's response to questions determines which page he should turn to next. If a question is not answered correctly he cannot move on to the next step.

Mathematics and Logic for Digital Devices by James T. Culbertson, Van Nostrand, 1958, is a textbook dealing with permutations and combinations, probability, number systems, traditional logic, Boolean algebra of classes, Boolean algebra of proportions, and applications to switching circuits. Ivan Flores' *Computer Logic: the Functional Design of Digital Computers*, Prentice-Hall, 1960, contains a useful annotated bibliography. Under the heading Boolean Logic, for example, *Engineering Applications of Boolean Algebra* by Boris Beizer and Stephen W. Leibholz (Gage, New York) is stated to be a most useful introduction, being written for the engineer who does not have previous logical background.

Two additional books which contain bibliographies are *Arithmetic Operations in Digital Computers* by R. K. Richards (Van Nostrand, 1955), and *Mathematics and Computers* by George R. Stibitz and Jules A. Larrivee (McGraw-Hill, 1957).

Books on numerical analysis are grouped in the Dewey class 517.6. Here will be found such works as *Mathematical Methods for Digital Computers* edited by Anthony Ralston and Herbert S. Wilf, Wiley, New York, 1960. In its preface it is described as "a reference text for many of the more commonly used mathematical methods for digital computers". The volume is in six parts, dealing respectively with the generation of elementary functions; matrices and linear equations; ordinary differential equations; partial differential equations; statistics; and miscellaneous methods. Somewhat similar ground is covered in *Numerical Methods for High Speed Computers* by G. N. Lance, Iliffe, London, 1960. An excellent, annotated bibliography of over 200 items is included in the National Physical Laboratory's *Modern Computing Methods* (*Notes on Applied Science, No.* 16), H.M.S.O., London, 2nd edition, 1961. For those requiring a textbook for secondary school and early college level, there is R. Wooldridge's *An Introduction to Computing*, Oxford University Press, London, 1962.

The above are, of course, only representative of a very extensive literature. Some of the works quoted include guides to further reading, and new contributions are reviewed in the periodicals described in the following section. Before leaving books, however, computer programming should be mentioned. A short list must suffice to illustrate the literature of this subject:

BOOTH, KATHLEEN H. W., *Programming for an Automatic Digital Calculator*. Butterworth, London, 1958.

EVANS, G. W. and PERRY, C. L., *Programming and Coding for Automatic Digital Computers*. McGraw-Hill, New York, 1961.

JEENEL, JOACHIM, *Programming for Digital Computers*. McGraw-Hill, New York, 1959.

NATHAN, ROBERT and HANES, ELIZABETH, *Computer Programming Handbook: a Guide for Beginners*. Prentice-Hall, Princeton, New Jersey, 1961.

WRUBEL, MARSHALL H., *A Primer of Programming for Digital Computers*. McGraw-Hill, New York, 1959.

Periodicals and Abstracts

The latest results of research in the field of computers are published in periodicals and technical reports. *United States Government Research Reports* are particularly valuable sources of information. To quote an example, UCRL–6581, 1961, by Rose Kraft and Carl J. Wensrich, is entitled *Monte Carlo Methods. A Bibliography Covering the Period 1949 to June 1961*. The procedure for identifying further relevant titles is set out in Appendix 2.

Mathematics of Computation is published quarterly by the

American Mathematical Society for the National Academy of Sciences—National Research Council. The topics which it covers include advances in numerical analysis, the application of numerical methods and high-speed calculator devices, the computation of mathematical tables, and the theory of high-speed calculating devices and other aids to computation. It also contains reviews and notes.

Another valuable American publication is *Communications of the Association for Computing Machinery* which contains papers dealing with both fundamental research and with computer applications.

In Britain, the *Computer Journal*, published by the British Computer Society, has been established for several years and is an acknowledged source of information of high quality.

There are several other titles which contain a significant amount of fundamental information. These include: *Avtomatika i Telemekhanika* (Russian, available in English translation), *Chiffres* (French), *I.R.E. Transactions*, *Journal of Mathematics and Physics*, *Journal of the Association for Computing Machinery*, *Numerische Mathematik* (German), and *Quarterly of Applied Mathematics*. *Computers and Automation* is particularly noteworthy for its reviews of equipment and services and for its *Computer Directory*.

A convenient way of discovering which periodicals include relevant information is to note the titles which regularly appear in the references contained in the two complementary abstracting publications which cover the field. These are *Computer Abstracts*, Technical Information Company, London, and *Computing Reviews* published bi-monthly by the Association for Computing Machinery. The former is issued monthly and includes relatively brief abstracts, while the latter's entries are more selective and evaluative. Although these two are specifically computer-orientated, the relevant sections of the invaluable *Mathematical Reviews* should not be left out of account.

Computer Societies

Evidence of the recognition of the necessity for an efficient system of recording and disseminating newly acquired information is further provided by the rapid growth of computer societies. Right from the start, national organizations have been keen to co-operate, to pool their human resources. This spirit can be seen in the foundation of the Provisional International Computation Centre, in the co-operative effort behind *Computing Reviews*, and in various other reciprocal arrangements between individual societies. Exchanges of information between East and West have undoubtedly led to a more rapid advancement of the *state of the art* than would otherwise have been the case.

It is essential that information seekers recognize the value of scientific and technical societies. Their interests and activities embrace every aspect of the fields they cover. One has only to peruse a year's back file of a computer society's journal to appreciate this. From its book reviews and advertisements one can compile a bibliography; from an examination of the titles of its principal articles one can grasp the range of computer applications; from its details of meetings one can assess the respective importance of areas of current research.

The Provisional International Computation Centre (Palazzo degli Uffici Zona dell EUR, Rome, Italy) has stated that one of its aims is to undertake mathematical research. Its *Bulletin* reports on the activities of members as well as the Centre itself. Particulars of national computer societies are available from the Centre. Activities in Great Britain and the United States are based on:

> The British Computer Society,
> Finsbury Court, Finsbury Pavement,
> London, E.C.2.

and

> The Association for Computing Machinery,
> Mt. Royal and Guilford Avenues,
> Baltimore 2, Maryland.

Mathematical Linguistics and Machine Translation

Computers provide one of the links between the mathematician and the linguist. Although the subject of machine translation may more properly be considered a part of linguistics, and therefore belonging to the Dewey 400 class, the subject is of growing interest to mathematicians. A brief guide to some sources of information is therefore appropriate here.

One of the barriers to the free interchange of information has been the language problem. English, French and German have traditionally been the principal modern languages in which scientists have recorded their discoveries and technologists their practical accomplishments. Who can say which will be the most used languages of the future? Russian, Chinese and Japanese are certainly claiming an increasingly large measure of the attention of research workers. The launching of the first sputnik by the Soviet Union signalled the start of a gigantic campaign in Western countries to make Russian literature more readily available to English-speaking scientists. The governments of Britain and the United States have added weighty support to programmes of abstracting and translation, and professional societies have undertaken an unprecedented amount of work of this nature. Scientists are being encouraged to learn foreign languages and special courses have been designed by educational institutions to cater for them. There has been a spate of textbooks tailored to the needs of the busy scientist.

The language problem still remains, and efforts are being made to harness machines to the task of solving it. Machine translation is very much a co-operative venture. In an introduction to the *Proceedings of the (1960) National Symposium on Machine Translation* (Prentice-Hall, 1961), H. P. Edmunson remarked that "These interdisciplinary investigations have required the researches of linguists, mathematicians, and computer scientists".

That mathematicians and linguists are equally concerned in the application of mathematics to language problems in general is evidenced by the origin of relevant literature. The *Proceedings of the Eighth International Congress of Linguists* (Oslo University Press, 1958), for example, contains one paper by Joshua Whatmough entitled: "Mathematical linguistics", and another by Paul L. Garvin on machine translation. On the other hand, Volume 12 of the American Mathematical Society's *Proceedings of Symposia in Applied Mathematics* is concerned with the "Structure of language and its mathematical aspects".

It would be wrong to infer that it is only Western countries which have a stake in this field. A great deal of work in mathematical linguistics is being carried out in Russia, particularly at the University of Leningrad. An article by N. D. Andreyev of that institution, entitled "Models as a tool in the development of linguistic theory", appeared in the April–August issue of *Word* (**18**, 186–97). Pioneer work in machine translation has been done in Russia by D. Yu Panov, whose book *Automatic Translation* was published in 'English in 1960 by Pergamon Press. Several useful bibliographies have been produced, including:

DELAVENAY, EMILE, *An Introduction to Machine Translation*. Thames and Hudson, London, 1960.
A comprehensive study dealing with all aspects of the subject. It includes a bibliography and glossary.
DELAVENAY, E. and K., *Bibliography of Mechanical Translation*. Mouton, The Hague, 1960.
OETTINGER, A. G., *Bibliography of Mathematical Linguistics*. Harvard Computation Laboratory, 1957 (mimeographed).

The Office of Technical Services has also published a selected bibliography of some 50 book reviews and reports on machine translation that have appeared in *Technical Translations* and *U.S. Government Research Reports*. In 1954, the Massachusetts Institute of Technology commenced publication of a periodical called *Machine Translation* to report current information on the subject.

Mathematical Tables

General

Mathematical tables are tools designed to save the time and labour of those engaged in computing work. They are by no means new. The Greek, Claudius Ptolemy of Alexandria, gave a table of values of the chords of a circle at intervals of one half degree in his *Almagest*. Indeed, tables of a more elementary kind date from much further back in history. The origins and development of tables are dealt with in a convenient manner in the *Encyclopaedia Britannica*.

With the onset of the electronic computer era the rate at which new tables are produced has been enormously accelerated, particularly in Great Britain, the United States, and the U.S.S.R. The problem now is to keep track of them so as to ensure their maximum use, and to avoid duplication of effort. To this end the National Research Council in the United States commenced publication of a quarterly journal entitled *Mathematical Tables and other Aids to Computation*. Its aim was to help scientists keep abreast of new tabular computations. This is now issued by the American Mathematical Society under the title *Mathematics of Computation*.

United States

In the United States much valuable work was done by the Mathematical Tables Project which was financed by the Works Projects Administration of the City of New York. The Project was eventually absorbed into the Applied Mathematics Division of the National Bureau of Standards, which had been organized in Washington D.C. in July 1947.

The Computation Laboratory of the National Bureau of Standards is the centre of government activity today. Its *MT* series of publications are important contributions to the subject. They have appeared in a number of publications. Many, for example, have been published in the *Journal of Mathematical Physics* and the *Bulletin of the American Mathematical Society*.

United Kingdom

An important series are the *British Association Mathematical Tables*. The series began publication in 1931. Ten volumes were published. In 1948 the Royal Society undertook responsibility for the work on mathematical tabulation and has published revised editions of some of the B.A. volumes. Any subsequent revisions and new volumes will be issued by the Royal Society. The Royal Society also publishes its own *Royal Society Mathematical Tables*, and the *Royal Society Shorter Mathematical Tables*.

The object of the *National Physical Laboratory Mathematical Tables Series* is to make available tables of mathematical functions prepared by the Mathematics Division of the N.P.L. which are likely to have a wide application but may not come within the range of the more fundamental tables published in the *Royal Society Mathematical Tables* series. One of the duties of the Mathematics Division is the construction of new mathematical tables and the revision of existing ones. Examples of this work are the tables contributed by F. W. J. Olver to the *Journal of Applied Physics* (1946) and the assistance given in the calculation and arrangement of the mathematical tables in Kaye and Laby's *Tables of Physical and Chemical Constants*. The Division constructs new tables in response to requests from clients, and some of these have been published in scientific journals. Others have been contributed to the depository for unpublished tables (from which photocopies may be obtained) which is maintained by the Royal Society Mathematical Tables Committee.

L. Fox's *The Use and Construction of Mathematical Tables*, 1956, is the first volume of the N.P.L. series. The subject is arranged under three headings: the use of mathematical tables, derivation of formulae and analysis of error, and the construction of mathematical tables. A list of references at the end of the volume constitutes a useful bibliography.

Guides to Tables

Indexes of tables help to avoid duplication and reduce the work involved in compiling new ones. They also show where gaps exist. Some of the more important, in chronological order of publication, are:

LEHMER, D. H., *Guide to Tables in the Theory of Numbers*, National Research Council, Washington, D.C., 1940.

BATEMAN, H. and ARCHIBALD, R. C., "Guide to tables of Bessel functions", *Mathematical Tables and Other Aids to Computation*. No. 7,1944.

FLETCHER, A. *et al.*, *Index of Mathematical Tables*, Addison-Wesley, Reading, 2nd edition, 1961.
This was first published by Scientific Computing Service Ltd., London, in 1946. The new edition (in two volumes) is divided into three parts.
Part 1 is arranged according to mathematical function and contains an extensive index.
Part 2 lists several thousand references to tables.
Part 3 lists known errors in published tables.

SCHÜTTE, KARL, *Index of Mathematical Tables*, Oldenburg, Munich, 1955.
Throughout this German work all textual matter, including headings and subheadings, is given in both German and English. The work is divided into 16 chapters and over 130 subdivisions. Within these the tables are arranged chronologically. Some 1200 tables are listed. There is an author index and an alphabetical list of institutions.

LEBEDEV, A. V. and FEDOROVA, R. M., *Guide to Mathematical Tables*, Pergamon Press, 1960.
This is a translation of a work compiled for the Computing Centre of the Academy of Sciences of the U.S.S.R. It is in two parts, of which the first, comprising 15 chapters, is arranged by mathematical function or type of operation. The items in this part bear reference numbers to the second part which contains bibliographical descriptions of the tables. Coverage is international.
—Supplement No. 1, by N. M. Burunova, Pergamon Press, 1960, contains information on tables published since the compilation of the main work, and other tables which had previously escaped the notice of the compilers. The arrangement is substantially the same as that of the parent work.

Collections of Tables

There are many useful collections of mathematical tables designed for different classes of users. The following are given as examples of recent publications of this type:

Barlow's Tables of Squares, Cubes, Square Roots, Cube Roots and Reciprocals of all Integers up to 12,500, 4th edition by L. J. Comrie, Chemical Publishing Co., New York, 1956.

BURINGTON, RICHARD STEVENS, *Handbook of Mathematical Tables and Formulas*, Handbook Publishers, Sandusky, Ohio, 3rd edition, 1949 (repr. 1958).
Contains 30 standard tables and 7 sections of formulas.

DWIGHT, HERBERT BRISTOL, *Tables of Integrals and Other Mathematical Data*, Macmillan, New York, 3rd edition, 1957.

JAHNKE, E. *et al.*, *Tables of Higher Functions*, McGraw-Hill, New York, 6th edition, 1960.

RYSHIK, I. M. and GRADSTEIN, I. S., *Tables of Series, Products and Integrals*, VEB Deutscher Verlag der Wissenschaften, Berlin, 1957.
Translation of a standard Russian work containing indefinite and definite integrals as well as Fourier and Laplace transformations. These are supplemented by some numerical tables. There is also a very good bibliography.

Logarithmic tables occur in a variety of forms and prices. One of the variants is the number of decimal places to which they have been worked. Chambers, for example, have catered for different requirements in this respect. A recent, compact volume issued by this publisher is

Chambers's Seven-Figure Mathematical Tables: Full Edition, compiled by James Pryde, London, 1961.

Mention of two further works must suffice as illustrations. They are, firstly:

Logarithmetica Britannica: being a Standard Table of Logarithms to Twenty Decimal Places, by Alexander John Thompson.
This is a publication of the Department of Statistics of University College, London, issued by Cambridge University Press. Nine separate sections have been published, and the complete work, consisting of logarithms of numbers 10,000–100,000, is available in two bound volumes. Cambridge University Press also issues the Department's series of *Tracts for Computers*.

and secondly:

Smithsonian Logarithmic Tables to Base e and Base 10, Smithsonian Institution, Washington, D.C., 1952.

Exercises

1. How are the various aspects of computers and computation treated in the Dewey Decimal Classification?
2. How would you decide which periodicals would be particularly useful to a mathematician in keeping abreast of new developments in the computer field?
3. Describe, with examples, the different types of sources of information on mathematical tables.
4. Write notes on the value of mathematical tables.

CHAPTER 8

Mathematical History and Biography

510.9: History of mathematics

Introduction

It is impossible to separate the history and biography of mathematics. The history of the subject is largely an account of the work of individuals. It should, however, be noted that in most libraries, biographies of mathematicians will be classified in the Biography class at the number 925.1. For libraries specializing in mathematics, provision is made in the schedules of the Dewey Decimal Classification for placing biography at 510·92.

Poggendorff

Deserving special mention is the monumental *Biographisch–literarisches Handwörterbuch der exakten Naturwissenschaften* of J. C. Poggendorff. It is a systematic bio-bibliographical reference work covering scientists of many nations. Following the biography of each individual is a full listing of his writings. The current volume, which is being issued in parts, began publication by Akademie Verlag of Berlin in 1955. Earlier volumes were issued by Barth of Leipzig in the following order:

1 ⎫	Origins–1858 (1864)	4	1883–1904 (1904)
2 ⎭		5	1904–22 (1925)
3	1858–83 (1898)	9	1923–31 (1936–40)

The years in brackets indicate the dates of publication. Poggendorff will not, of course, be found in all libraries, but all serious students of mathematics should be aware of its existence.

General Histories

Articles in the better encyclopedias are very useful for scene-setting and orientation. Articles on mathematical history may be supplemented by others dealing with the lives and works of individual mathematicians. Usually they also provide guides to further reading.

Similarly, the well-known histories of science may be profitably used, particularly for seeing the development of mathematics against a wider scientific background. Two such works are Charles Singer's *A Short History of Scientific Ideas* (Oxford University Press, 1957), and George Sarton's *A History of Science* in two volumes (Harvard University Press, 1952–9).

General histories of mathematics vary in their treatment of the subject. Some give a straightforward chronological picture from the earliest times to the present, others concentrate on a particular period, and a few offer selections of original writings by eminent mathematicians.

High on one's reading list should be *The Development of Mathematics* by Eric Temple Bell (McGraw-Hill, New York, 2nd edition, 1945), and the same author's *Mathematics, Queen and Servant of Science* (McGraw-Hill, 1951). Lancelot Hogben's *Mathematics in the Making* (Macdonald, London, 1960) is a beautifully illustrated account of the development of mathematics which breaks away from the style of the more formal history text. Hogben is perhaps best known for his popular *Mathematics for the Million*.

There are several other comprehensive histories of note which are well represented in libraries. They include:

BALL, W. W. ROUSE, *Short Account of the History of Mathematics*, Dover, New York, 1960.
This was previously published by Macmillan, 3rd edition, 1901.
CAJORI, FLORIAN A., *History of Mathematics*, Macmillan, New York, 2nd edition, 1919 (repr. 1953).
FREEBURY, H. A., *History of Mathematics for Secondary Schools*, Cassell, London, 1958.
SMITH, DAVID EUGENE, *History of Mathematics*, 2 volumes, Dover, New York, 1958. (Originally published by Ginn, Boston, 1923–5.)

Vol. 1: *General Survey of the History of Elementary Mathematics.*
   The chapters are arranged chronologically and are preceded by a
   selective bibliography.
Vol. 2: *Special Topics of Elementary Mathematics.*
STRUIK, DIRK JAN, *A Concise History of Mathematics*, Bell, London,
   1954.

Morris Kline has been concerned to show the relationship of
mathematics to general cultural development. His latest work,
Mathematics: a Cultural Approach, was published in 1962 by
Addison-Wesley, Reading, Mass., and London. Two other
contributions by Kline are *Mathematics and the Physical World*
(Crowell, New York, 1959), and *Mathematics in Western Culture*
(Oxford University Press, New York, 1953).

Selections from 125 different, important mathematical writings
are presented in David Eugene Smith's *A Source Book in Mathematics*, 2 volumes, Dover, New York, 1959. This Dover edition
is stated to be an unabridged and unaltered republication of the
first edition originally published as a one-volume work in 1929.
The extracts, translated into English where necessary, are grouped
under broad headings: the field of number, algebra, geometry,
probability, calculus, functions, and quaternions.

Special Periods

Whilst the general histories usually recount the story from the
early beginnings of mathematics up to the present, other works
deal selectively with particular periods of development.

Otto Neugebauer's *The Exact Sciences in Antiquity* (Brown
University Press, Providence, R.I., 2nd edition, 1957) includes
Babylonian, Egyptian and Greek manuscripts. With A. J. Sachs,
Neugebauer also wrote *Mathematical Cuneiform Texts* (American
Oriental Society, New Haven, Conn., 1945) relating to mathematics
in Mesopotamia.

Greek mathematics is covered by Sir Thomas Little Heath's
A History of Greek Mathematics, 2 volumes, Oxford University
Press, 1921. This is a detailed work for the advanced student. A

more recent volume is that by Tobias Dantzig entitled *The Bequest of the Greeks*, Scribner, New York, 1955.

The 17th and 18th Centuries figure in *Classical Mathematics, a Concise History of the Classical Era in Mathematics* by Joseph Ehrenfried Hofmann (Philosophical Library, New York, 1959).

Biography

The outstanding single work under this heading is Poggendorff's handbook which was described at the beginning of this chapter.

Detailed studies of individual mathematicians have also appeared, of which J. F. Scott's *The Mathematical Work of John Wallis, D.D., F.R.S. (1616–1703)* may be taken as an example. Originally presented as a doctoral thesis, this was published by Taylor and Francis, London, in 1938. One appendix gives brief biographies of Wallis's contemporaries and immediate predecessors, and another lists the subject's mathematical writings, including his contributions to the *Transactions of the Royal Society*.

Two series of memoirs may be noted as examples of another useful type of source material. They are the annual *Biographical Memoirs of Fellows of the Royal Society*, in which each entry is followed by an exhaustive bibliography of the subject's writings arranged chronologically, and the *Biographical Memoirs* of the National Academy of Sciences of the U.S.A. which is published annually by Columbia University Press. An example from the latter is an article on Eliakim Hastings Moore, 1862–1932, contributed by G. A. Bliss and L. E. Dickson (**17**, 83–102, 1937). The piece is followed by a bibliography of 75 references to Moore's writings.

Whilst Smith's source book reproduces extracts from the writings of many eminent mathematicians, the *Collected Works of John Von Neumann*, edited by A. H. Taub (Pergamon Press, 1961), covers the output of a single man in great detail.

In order to keep abreast of new publications on mathematical history and biography, the appropriate section of *Mathematical Reviews* should be regularly perused. Articles on historical research in mathematics are included in such periodicals as the new *Archive for History of Exact Sciences* (1960–) published by Lange and Springer, Berlin–Wilmersdorf, Germany.

Mathematicians of Today

Those actively engaged in mathematics usually belong to one or more of the national societies, and the simplest way of identifying them is through society membership directories. The libraries of most mathematical societies usually have a collection of membership lists of corresponding overseas organizations, whilst the larger public libraries are also able to produce a number of them, particularly those which appear in issues of a society's periodical publication. Individual membership lists which may be specifically required are available direct from the societies concerned. Names and addresses of societies will be found both in Chapter 5 and in chapters dealing with special branches and aspects of mathematics. An excellent example is the *Combined Membership List* of the Mathematical Association of America, the Society for Industrial and Applied Mathematics, and the American Mathematical Society. It costs $5.00, and is regularly revised.

The most up-to-date biographical notes of all are to be found in the news or personalia sections of mathematical periodicals.

A service offered by some societies is their membership mailing list which can be supplied, for purposes in the interest of members, either in a form convenient for affixing to envelopes or already printed on a set of envelopes. The American Mathematical Society's list comprises some 8500 names.

A *World Directory of Mathematicians* has been published in Bombay by the Tata Institute of Fundamental Research. It is now in its second edition, 1961.

The H. W. Wilson Company of New York publishes *Current Biography* each month (except August). It covers all fields and has a yearly cumulation known as *Current Biography Yearbook*. Each entry is followed by a list of references to the sources of biographical information used. A criterion for inclusion in this publication is that the person must be "prominent in the news". A recent example is the Russian mathematician Mstislav Keldysh (Feb. 1962, pp. 19–21). Obituary notices are published for persons whose biographies have previously appeared.

Eminent mathematicians are represented in the various who's who publications, and certainly appear in *American Men of Science* (Cattell, Tempe, Arizona, 10th edition, 1960–62). This is a 5 volume work which gives biographical data on living Americans, including Canadians.

Mathematicians working in academic institutions may be identified in prospectuses and calendars, together with details of any research projects they are engaged upon. *Research in British Universities*, issued annually on behalf of the Department of Scientific and Industrial Research by H.M. Stationery Office, London, is a comprehensive British source.

Another way of locating individuals working on particular problems is through the publishers of journals in which their papers have appeared. Occasionally, publishers give the addresses of contributors specifically for this purpose.

Exercises

1. Discuss the question of whether biographical works should be classified with the appropriate subjects or grouped together under the general heading, Biography.
2. How would you set about tracing the mathematical contributions of any living British or American mathematician?
3. Write notes on the relative importance to the student of mathematical history, of histories of mathematics and the original writings of eminent mathematicians.

Illustrations

The following reproductions are illustrative of the type of information which will be found in some of the sources described in the text.

MATH'E-MAT'I-CAL, *adj.* **mathematical expectation.** See EXPECTATION.

mathematical induction. See INDUCTION.

MATH'E-MAT'ICS, *n.* The logical study of shape, arrangement, and quantity.

applied mathematics. A branch of mathematics concerned with the study of the physical, biological, and sociological worlds. It includes mechanics of rigid and deformable bodies (elasticity, plasticity, mechanics of fluids), theory of electricity and magnetism, relativity, theory of potential, thermodynamics, biomathematics, and statistics. Broadly speaking, a mathematical structure utilizing, in addition to the purely mathematical concepts of space and number, the notions of time and matter belongs to the domain of applied mathematics. In a restricted sense, the term refers to the use of mathematical principles as tools in the fields of physics, chemistry, engineering, biology, and social studies.

mathematics of finance. The study of the mathematical practices in brokerage, banking, and insurance. *Syn.* Mathematics of investment.

pure mathematics. The study and development of the principles of mathematics as such (for their own sake and possible future usefulness) rather than for their immediate usefulness. *Syn.* Abstract mathematics. See above, applied mathematics.

MATHIEU'. **Mathieu's differential equation.** A differential equation of the form

$$y'' + (a + b \cos 2x)y = 0.$$

The general solution can be written in the form $y = Ae''\phi(x) + Be^{-rb}(-x)$, for some constant r and function $\phi(x)$ which is periodic with period 2π. There are periodic solutions for some *characteristic values* of a, but no Mathieu equation (with $b \ne 0$) can have two independent periodic solutions.

Mathieu function. Any solution of Mathieu's differential equation which is periodic and is either an even or an odd function, the solution being multiplied by an appropriate constant. The solution which reduces to $\cos nx$ when $b = 0$ and $a = n^2$, and for which the coefficient of $\cos nx$ in its Fourier expansion is unity, is denoted by $ce_n(x)$; the solution which reduces to $\sin nx$ when $b \to 0$, and in which

the coefficient of $\sin nx$ in its Fourier expansion is unity, is denoted by $se_n(x)$.

MA'TRIX, *adj.*, *n.* [*pl.* **matrices**]. A rectangular array of terms called **elements** (written between parentheses or double lines on either side of the array), as

$$\begin{pmatrix} a_1 & b_1 & c_1 \\ a_2 & b_2 & c_2 \end{pmatrix} \quad \text{or} \quad \begin{Vmatrix} a_1 & b_1 & c_1 \\ a_2 & b_2 & c_2 \end{Vmatrix}.$$

Used to facilitate the study of problems in which the relation between these elements is fundamental, as in the study of the existence of solutions of simultaneous linear equations. Unlike determinants, a matrix does not have quantitative value. It is not the symbolic representation of some polynomial, as is a determinant (see below, rank of a matrix). If the elements of a matrix are all real, the matrix is a **real matrix.** A **square matrix** is a matrix for which the number of rows is equal to the number of columns. The number of rows (or columns) is called the **order** of the matrix. The diagonal from the upper left corner to the lower right corner is called the **principal** (or **main**) **diagonal.** The diagonal from the lower left corner to the upper right corner is called the **secondary diagonal.** The **determinant** of a square matrix is the determinant gotten by considering the array of elements in the matrix as a determinant. A square matrix is **singular** or **nonsingular** according as the determinant of the matrix is zero or nonzero. A **diagonal matrix** is a square matrix all of whose nonzero elements are in the principal diagonal. If, in addition, all the diagonal elements are equal, the matrix is a **scalar matrix.** An **identity** (or **unit**) **matrix** is a diagonal matrix whose elements in the principal diagonal are all unity. For any square matrix A of the same order as I, $IA = AI = A$.

adjoint of a matrix. See ADJOINT.

associate matrix. See HERMITIAN -Hermitian conjugate of a matrix.

augmented matrix of a set of simultaneous linear equations. The matrix of the coefficients, with an added column consisting of the constant terms of the equations. The augmented matrix of

$$\begin{matrix} a_1x + b_1y + c_1z + d_1 = 0 \\ a_2x + b_2y + c_2z + d_2 = 0 \end{matrix} \text{ is } \begin{Vmatrix} a_1 & b_1 & c_1 & d_1 \\ a_2 & b_2 & c_2 & d_2 \end{Vmatrix}.$$

FIG. 1 James & James *Mathematics Dictionary* Copyright 1959, D. Van Nostrand & Co., Inc.

MATHEMATICS

General

770a. GREEN, R. A. C. (Lei). Many-valued logics: a study of the relationship of propositional calculi and algebraic systems. PH.D.

See also 759, 3189.

Theory of Numbers

771. DOWIDAR, A. F. El-S.A. (W, *Swansea*). Some problems on number theory and analysis. PH.D.

Algebra

772. HAYES, A. (C, *Trinity*). Representations of partially-ordered algebraic systems. PH.D.
773. HALLETT, J. T. (LQMC). Torsion-free abelian groups whose automorphism groups are finite. PH.D.
774. LAXTON, R. R. (LKC). Near-rings with descending chain conditions. PH.D.
775. MOHAMED, I. J. (LQMC). On series of subgroups related to groups of automorphisms. PH.D.
776. AUSTIN, A. K. (M). Finite metacyclic groups. M.Sc.
777. BULL, R. A. (M). Some axioms for varieties of groups. M.Sc.
778. DICKER, R. M. (O, *Balliol*). Topics in abstract algebra. D.PHIL.
779. HARON, A. E. P. (O, *Magdalen*). Problems in homological algebra. D.PHIL.
780. HOARE, A. H. M. (O, *Trinity*). Some group theoretical problems. D.PHIL.
781. TEH, H. H. (Q). Extensions and representations of partially ordered groups. M.Sc.
782. DAYKIN, D. E. (R). Hilbert's 17th and other problems. PH.D.
783. THOMPSON, A. C. (R). The theory of general systems of linear equations. M.Sc.

42

FIG. 2 "Vast sums of money are invested in university research activities, and one way in which the dividends may be increased is by more efficient use of doctoral theses". (p. 35).
Index to Theses Accepted for Higher Degrees in the Universities of Great Britain and Ireland, Volume XI.
By kind permission of Aslib.

TEXAS Oil Jobber. See PETROLEUM

VARIETY Store Merchandiser. (Prints a different section
each month) 1931. m. $6.50. Ed. Preston J. Beil. Vari-
ety Store Merchandiser Publications. 419 Park Ave. S.,
New York 16, N.Y. adv. illus. mkt. tr.lit. circ. 27,000-.

WESTERN Advertising; the magazine of advertising and
marketing for the thirteen Western States and British
Columbia. See ADVERTISING

MATHEMATICS

ACADÉMIE Polonaise des Sciences. Bulletin. Serie des Sciences Mathématiques, Astronomique et Physiques. 1953.
irreg. 20 zł. Ed. A. Mostowski. Miodowa 10, Warsaw,
Poland. Indexed: Chem.Abstr.

ACADEMY of Sciences of the USSR. Bulletin. Mathematical
Series. (English translation of "Izvestiya Akademiya" Nauk
SSSR. Seria Matematika) 1962. bi-m. Rs.200.($60.) Ed.Bd.
Hindustan Publishing Corp.,(1) 6 U.B. Jawahar Nagar, Delhi,
India. bibl. bk.rev. charts. illus. tr.lit. index.
circ. 800 approx.

ACCADEMIA Nazionale dei Lincei. Classe di Scienze Fisiche,
Matematiche e Naturali. Rendiconti. See PHYSICS

ACTA Mathematica. 1882. irreg. Kr.80. Ed. N.E. Norlund.
Institut Mittag-Leffler, Auravagen 17, Djursholm 1, Sweden.
bibl. index. cum.index: v.1-100. Indexed: Math.R.

ACTA Mathematica. (Academiae Scientiarum Hungaricae)
(Text in English, French, German and Russian) 1950.
s-a.(2 double nos. to a vol) $8.50 per vol. Ed. G. Hajos.
Akadémiai Kiado, Publishing House of the Hungarian Academy of Sciences, Alkotmány u.21, Budapest V, Hungary.
adv. bibl. bk.rev. index. Indexed: Math.R.

ACTA Polytechnica Scandinavica. Mathematics and Computing
Machinery Series. See ENGINEERING

FIG. 3 *Ulrich's Periodicals Directory* (Eileen C. Graves, ed.). 10th
edition, 1963 (see p. 39). By kind permission of R. R. Bowker
Company.

Mathematical Reviews

Vol. 25, No. 1   January 1963   Reviews 1–1090

GENERAL

★**American Mathematical Society Translations.**   **1**
  Series 2, Vol. 20: 6 papers on partial differential equations.
  American Mathematical Society, Providence, R.I., 1962.
  iii + 364 pp. $5.30.
This volume contains translations of articles of B. M. Levitan [#317], M. A. Naĭmark [#464], O. A. Ladyženskaja [#342], A. I. Košeler [#323], E. M. Landis [#316], and M. I. Višik and L. A. Ljusternik [#322].

Obádovics, J. Gy.   **2**
  ★**Taschenbuch der Elementar-Mathematik. Mit praktischen Anwendungen.**
  I. Auflage der deutschen Ausgabe nach der III. umgearbeiteten und erweiterten ungarischen Auflage. Aus dem Ungarischen übersetzt von Dr. Johanna Raab, István Miklos.
  Akadémiai Kiadó, Budapest, 1962. 868 pp. $4.80.
A translation into German of the third Hungarian edition [Akad. Kiadó, Budapest, 1961]. The handbook contains chapters on arithmetic, algebra, geometry, vector algebra, differential and integral calculus, ordinary differential equations.

List of books and papers by Teiji Takagi   **3**
  (1875–1960). (Japanese)
  Sûgaku 12 (1960/61), 135–136.

HISTORY AND BIOGRAPHY

Berezkina, E. I.   **4**
  Le traité mathématique de Sun-zi Suan-Jing. (Russian)
  Istor.-Mat. Issled. 13 (1960), 219–230.
This is a description of a largely arithmetical anonymous Chinese treatise composed in the third or fourth century of our era. It consists of three "books", of which the first includes, among other numerical tables, a tabulation of m^2n^2 and mn^2 for $n=9, 8, 7, \cdots, 1$; $m=n, n-1, n-2, \cdots, 1$. It also has a short description of the use of the computing board, and two systems for designating high powers of ten. The second and third books are made up of sixty-four worked examples involving proportions, percentages, progressions, metrology, areas of fields, square roots, systems of linear equations, and so on. Of particular interest is an early example of linear indeterminate equations. The approximation rule $\sqrt{(a^2+r)} \approx a + r/2a$ is

given. The Sun-zi is useful historically in that the language of the demonstrations is less laconic than that of the earlier and more systematic "Arithmetic in Nine Books", thus permitting fuller insight into the methods of solution.
              E. S. Kennedy (Princeton, N.J.)

Zubov, V. P.   **5**
  Le traité "De continuo" de Bradwardine. (Russian)
  Istor.-Mat. Issled. 13 (1960), 385–440.
This treatise by Thomas Bradwardine, the Oxford scholar, written between 1328 and 1335, is here published in the original Latin after the two extant manuscripts, that of Thorn (Torun), described in detail by M. Curtze [Z. Math. Phys. 13 (1868), supplement, 45–104], and that of Erfurt, mentioned by W. Schum [*Beschreibendes Verzeichnis der Amplonianischen Handschriften-Sammlung zu Erfurt*, Berlin, 1887, pp. 641–642]. The initial page of each manuscript is also reproduced, and an outline is given (pp. 385–429) of the text. This work of Bradwardine has also been discussed by J. E. Murdoch [Actes du IXᵉ Congrès Intern. d'Hist. des Sciences (Barcelona-Madrid, 1959), pp. 538–543].   *D. J. Struik* (Cambridge, Mass.)

LOGIC AND FOUNDATIONS

See also 35, 37, 139, 140, 774, 776, 777, 778, 779, 1088, 1089.

Quine, Willard Van Orman   **6**
  ★**Mathematical logic.**
  Revised edition. Harper Torchbooks: The Science Library.
  Harper & Row, Publishers, New York-Evanston, Ill., 1962. xii + 346 pp. $2.25.
A paper-bound reprint of the well-known revised edition of 1951 [Harvard Univ. Press, Cambridge, Mass., 1951; MR 13, 613].

Thomas, Ivo; Orth, Don   **7**
  Axioms for the "Gergonne"-relations.
  J. Symb. Logic 24 (1959), 305.
The authors investigate the independence of the axioms of a formal system closely related to the system of Faris [same J. 20 (1955), 207–231; MR 17, 701] for the "Gergonne"-relations. They replace Axiom 9, $ClabKN2abKN3abN5ab$ by the three axioms $9iClabNiab$ ($i=2, 3, 5$), and make a similar replacement for Axiom 10. Otherwise the formal system is unchanged. It is shown that Axiom 2 is probable from the remaining axioms, but

                        1

FIG. 4   *Mathematical Reviews* (see pp. 46–48).
By kind permission of the American Mathematical Society.

THE
MATHEMATICAL
GAZETTE

The Journal of the
Mathematical Association

Vol. XLVI No. 358   December 1962

FIG. 5 "Periodicals concerned with mathematics teaching should be
regularly perused in order to keep abreast of the latest techniques".
(p. 65).
Contents page of *The Mathematical Gazette*.
By kind permission of the Mathematical Association.

GEOMETRY—*cont.*

513.8—Non-Euclidean geometry
513.82—n-Dimensional geometry

KENDALL, Maurice George
A course in the geometry of n dimensions. London,
Griffin, 16/-. 1961 [i.e. Feb 1962]. *viii,63p. bibliog.
21½cm. Lp. (Statistical monographs and courses, edited
by M.G. Kendall—no.8)*

(B62-3881)

*513.82[1]—Banach spaces

HOFFMAN, Kenneth
Banach spaces of analytic functions. London, Prentice-
Hall, 50/-. Aug 1962. *xiii,217p. bibliog. 23½cm.*

(B62-12569)

513.82[1]—Linear spaces

AMIR-MOEZ, Ali Reza, and FASS, Arnold Lionel
Elements of linear spaces. Oxford, London, Pergamon
35/-. Jun 1962. *ix,149p. illus., diagrs. 29cm. (International series of monographs on pure and applied mathematics, edited by I. N. Sneddon, and others—vol.26)*

(B62-10414)

SHILOV, Georgi Evgen'evich
An introduction to the theory of linear spaces;
translated from the Russian by Richard A. Silverman.
London, Prentice-Hall, 50/-. 1961 [i.e. Feb 1962].
ix,310p. diagrs., bibliog. 24cm.

(B62-3882)

513.83—Topology
513.83ab—*Periodicals*

TOPOLOGY: an international journal of mathematics,
founded by J. H. C. Whitehead. Vol.1. Oxford, London,
Pergamon, £10 (70/- to individuals) per annum. Jan
1962. *25cm. Sd.*
Quarterly.

(B62-6383)

FIG. 6   "The *British National Bibliography* (B.N.B.) is arranged according
to the Dewey Decimal Classification. . . . First issues of new
periodicals and those which have changed their titles are also
noted". (see pp. 103–104).
British National Bibliography, Annual Volume 1962.
By kind permission of the Council of the B.N.B., Ltd.

Term	Symbol
function of x	$f(x)$, $F(x)$, etc.
increment or finite difference operator	Δ, δ
*differential coefficient of y with respect to x	$\dfrac{dy}{dx}$, dy/dx
*differential coefficient, n^{th}	$\dfrac{d^ny}{dx^n}$, d^ny/dx^n
†differential coefficient, partial	$\dfrac{\partial y}{\partial x}$, $\partial y/\partial x$
operator $\dfrac{d}{dx}$	D
*integral of y with respect to x:	
indefinite	$\int y\, dx$
from $x = a$ to $x = b$	$\int_a^b y\, dx$
around a closed contour	$\oint y\, dx$
complex operator $\surd(-1)$ (see also Electricity and Magnetism, schedule B,5)	i, j
real part of ( )	$\mathscr{R}$ ( ), Re( )
imaginary part of ( )	$\mathscr{J}$ ( ), Im( )
modulus of complex number $z \equiv x + iy$	$\|z\| \equiv \|x + iy\|$
argument of complex number z	$\arg z$
vector of magnitude A	$\mathbf{A}$
scalar product of $\mathbf{A}$ and $\mathbf{B}$	$\mathbf{A.B}$

Fig. 7 Selection of mathematical symbols from British Standard 1991, Part 1, 1954, *Letter Symbols, Signs and Abbreviations: General.* By kind permission of the British Standards Institution, 2 Park Street, London, W.1, from whom copies of the complete standard may be purchased.

CHAPTER 9

Mathematical Books. Part 1: Bibliographies

511: Arithmetic
512: Algebra
513–516: Geometry
517: Calculus

Introduction

Opinions about books as sources of information vary greatly. They are largely based on the type of book used during the individual's period of most intensive study, or in the practice of his chosen profession. The textbook naturally figures prominently in many minds. In fact, one's respect for books as working tools can be conditioned by the quality of the particular volumes used during organized courses of study. Those who use books in their occupations consider them in the light of their own experience. The travelling salesman might see them as directories, the research worker as technical reports, and the actuary as statistical handbooks. There are many different kinds of books, each filling a specific role which is primarily determined by the type of information required. It should, however, be noted that different people may well use the same book for different purposes. It is important to be aware of the variety of books, to discover the relationships that exist between them, and to know which to use in particular circumstances.

Finding the books which meet one's requirements at any given time involves three processes: *identification* (discovering appropriate lists of authors and titles), *acquisition* (locating copies of the actual books), and *evaluation* (assessing the ability of the individual books to yield the required information). Identification is a pre-requisite to the other processes, and normally presents the greatest difficulties. Its various aspects therefore occupy the whole of this chapter. Acquisition and evaluation are dealt with in Chapter 10.

Bibliographies

Problems of Identification

The need for books may be said to arise through the desire or necessity to find information on a particular subject, and it depends on the purpose for which the information is required what *kind* of books will be sought. For general orientation in a subject an encyclopedia article may suffice. Factual data may be found in handbooks and standard reference works. Even general engineering handbooks, for example, usually contain useful chapters on mathematical fundamentals. Monographs deal exhaustively with narrowly defined subjects, and normally contain guides to further reading. For consecutive study, the textbook is the ideal medium. The advanced textbook having a systematic layout and detailed index may also, of course, be used as a reference tool.

It will be appreciated that the title of a book does not necessarily indicate what kind of book it is. Whilst the most satisfactory way of selecting books is to examine them physically, this is clearly not always possible. One frequently needs to have some means of discovering the authors and titles of books which *may* be suitable. This requirement is met by bibliographies.

Functions of Bibliographies

Bibliographies are compilations of references to published (and sometimes unpublished) literature on a particular place, person or subject. They are produced as aids to book selection and literature searching, and include sufficient information about each item to enable the enquirer to specify his requirements accurately either to a bookseller or to a librarian. This information consists of author, title, publisher, date and place of publication. Additional particulars may be supplied such as size, a note on the subject content, whether a list of references is included, and whether the text is supplemented by illustrative material such as maps, photographs, diagrams, facsimiles, tables, and so forth. A bibliography may be issued serially on a continuing basis, or as a complete work. In the latter case, the date limits of material quoted are usually indicated.

Bibliographies are used as means of identification of required material. They indicate the existence of this material without necessarily indicating where it may be obtained. In this way they differ from library catalogues and booksellers' lists. It should, however, be noted that both the latter can constitute bibliographies, especially when they have a subject limitation. Thus, for example, the printed catalogue of a mathematical society's library would form a valuable bibliography of mathematics, as would a bookseller's list of his mathematical stock.

The nature of the information provided in each entry determines whether the bibliography is *descriptive* or *evaluative*. The *List of Books Suitable for Training Colleges* mentioned in the section on textbooks is an evaluative bibliography as it is graded according to user requirements and contains comments on the suitability of individual titles for particular purposes.

Guides to Bibliographies

Whilst it is essential to know the books dealing with one's own subject, this is not sufficient. Problems of finding information frequently occur in fields bordering one's own. This is especially so in mathematics. The mathematician often finds himself working as a specialist member of a team which may be engaged on a problem outside the range of his subject experience. It then becomes imperative for him to be able to find his way through the literature of the new subject.

The following sections contain descriptions of the principal mathematical bibliographies, and to that extent the present work is itself a guide to bibliographies. The titles listed below have been selected both for their value as guides to mathematical bibliographies and as aids to orientating the enquirer in related fields.

Columbia University's *Guide to the Literature of Science*, 2nd edition, 1957, was prepared for use in connection with courses in science literature by its School of Library Service. Gertrude Schutze's *Bibliography of Guides to the S–T–M* (Science, Technology, Medicine) *Literature* has proved useful enough to warrant the publication of a new edition. It was published in 1958. Another helpful list is Burns, Robert W., "Literature resources for the sciences and technologies: a bibliographical guide", *Special Libraries* **53**, 262–71, 1962.

For those who wish to pursue the matter further two large bibliographies of bibliographies may be mentioned:

> BESTERMAN, THEODORE, *A World Bibliography of Bibliographies and of Bibliographical Catalogues, Calendars, Abstracts, Digests, Indexes and the Like*, Scarecrow Press, New York, 3rd edition, repr. 1960.
> BOHATTA, HANNS and HODES, FRANZ, *Internationale Bibliographie der Bibliographien*, Klostermann, Frankfurt a.M., 1950.

Robert L. Collison's *Bibliographies, Subject and National: a Guide to their Contents, Arrangement and Use* (Crosby Lockwood, London, 2nd edition, 1962) is a less ambitious compilation, but it includes the main bibliographies of a wide range of subjects.

General Mathematical Bibliographies

The first bibliography of mathematics was probably Cornelius à Beughem's *Bibliographia Mathematica*, comprising some 3000 entries, and published in Amsterdam in 1685 and 1688. Other early bibliographies, many of them German, are listed in *Scientific Books, Libraries and Collectors* by John L. Thorton and R. I. J. Tully, Library Association, London, 2nd edition, 1962. Nathan Grier Parke's *Guide to the Literature of Mathematics and Physics Including Related Works on Engineering Science*, Dover, New York, 2nd edition, 1958, is divided into two parts: 1, General Considerations; and 2, The Literature. The first deals with the principles of reading and study, self-directed education, literature search, and periodicals. Part 2 is the bibliography proper. It contains over 5000 entries grouped under subject headings arranged alphabetically. The number of entries is in fact more than double that of the first edition (McGraw-Hill, 1947), and not surprisingly the emphasis is on applied mathematics. New headings include Actuarial Mathematics (7 entries), Games of Strategy (mathematics) (12 entries), Operations Research (3 entries), and Sets, Theory of (8 entries). Introductory remarks, sometimes including additional references, precede the bibliographical entries under each subject heading. Definitions of the subjects are accompanied by notes on some of the titles quoted. Full author and subject indexes are included. A large proportion of the work naturally deals with the literature of physics. Peripheral topics of a wide range also appear, among them: chemistry, mechanical and general engineering, and technical writing.

Volume 3 of *Les Sources du Travail Bibliographique* by L. N. Malclès, Minard, Paris, 1958, is entitled *Bibliographies Spécialisées*. Mathematical sciences are dealt with in detail on pp. 30–55, arranged in two parts as follows:

> *Partie générale*
> Histoire: mathématiques générales; recueils de tables et de formules; bibliographies retrospectives; bibliographies internationales courantes; principaux périodiques mathématiques.
> *Partie spéciale*
> Arithmétique et algèbre; théorie des nombres, des groupes, des ensembles; calcul numérique. Analyse: calcul infinitésimal; théorie des fonctions; calcul des variations. Géométrie. Calcul des probabilités; statistique mathématique; logique mathématique.

Special Mathematical Bibliographies

Many bibliographies of value to mathematicians have been produced to cover more specialized needs. They may deal with a particular branch of the subject, the mathematical publications of a particular country, the literature of a specified period, or the requirements of an identified class of user, textbooks for example. Those which have a geographical specification include:

> *Bibliografia Brasileira de Matemática e Física.* Instituto Brasileiro de Bibliografia e Documentação, 1956–, Volume 1, 1950–54.

CHIA KUEI TSAO, *Bibliography of Mathematics Published in Communist China during the Period 1949–1960.* American Mathematical Society, 1961.
1335 entries with over 1160 titles.

FORSYTHE, GEORGE ELMER, *Bibliography of Russian Mathematics Books.* Chelsea Pub. Co., New York, 1956.
French Bibliographical Digest. Series II, No. 14: *Science and Mathematics,* Part 1: *Pure Mathematics.* French Cultural Services of New York, 1955.

KARPINSKI, L. C., *Bibliography of Mathematical Works Printed in America Through 1850.* University of Michigan Press, 1940.

LA SALLE, J. and LEFSCHETZ, S., *Recent Soviet Contributions to Mathematics.* Macmillan, New York, 1962.
Unione Matematica Italiana. *Bibliografia Matematica Italiana.* Tip.Ed. Cremonese, 1950–.

Individual aspects and branches of mathematics—history, teaching, statistics, and so forth—are the subject of other chapters of the present work, each containing details of appropriate bibliographies. It is a useful exercise to create a personal file of references to bibliographies in special mathematical fields. This can be done on standard index cards, and the exercise is doubly effective if these are arranged according to their Dewey classification.

Special bibliographies are frequently met with in periodicals. Examples are:

GOODSTEIN, R. L., "Pure mathematics", *British Book News,* No. 221, 1–5, 1959.
A brief survey of books on pure mathematics published in Great Britain since 1948. It deals with textbooks and monographs.

HOUSEHOLDER, ALSTON S., "Bibliography on numerical analysis", *Association for Computing Machinery Journal,* **3,** 85–100, 1956.

It will be apparent that potentially useful bibliographies occur in publications that will not normally be scanned by mathematicians. A check can, however, be kept on them by regularly consulting the *Bibliographic Index: a Cumulative Bibliography of Bibliographies.* This notes bibliographies appearing in about 1500 different periodicals. It is published semi-annually, with annual and larger cumulations, by Wilson, New York.

The fact should not be overlooked that many books on special aspects of mathematics contain excellent select bibliographies. Monographs are a particular case. They are books which aim to treat a topic exhaustively, setting down as far as possible all that is known about it. When dealing with monographs of mathematical topics it is important to consider their publication dates, and to consult the abstracting publications for possible new contributions. Clearly, the topic under consideration must be very specific. Early examples in scientific literature concerned the life cycles of individual insects. An important mathematical series are the *Carus Mathematical Monographs* of the Mathematical Association of America which may be exemplified by *Irrational Numbers* by Ivan Morton Niven (No. 11, 1956). The American Mathematical Society publishes two series of monographs: *Mathematical Surveys*, and *Colloquium Publications*. The latter deal with advanced mathematical subjects such as orthogonal polynomials, algebraic topology, and point set theory. Other such series include the *Princeton Mathematical Series* (1939–), and the *Cambridge Tracts in Mathematics and Mathematical Physics* (1905–).

A unique kind of bibliography is Eleanora A. Baer's *Titles in Series*, published by Scarecrow Press, Washington, D.C. Individual volumes are listed under the title of their parent series. The series themselves are arranged alphabetically. Thus under *Carus Mathematical Monographs* appear the authors, titles and dates of publication of the single works. An alphabetical author and title index is provided. The first volume of *Titles in Series* was published in 1953. Volume 2, 1957, supplements this and includes new series published prior to January 1957. Volume 3, 1960, brings the coverage up to the end of December 1959. A directory of publishers is given in each volume, and the two later volumes also include a cumulative index to serial titles.

Bibliographies of Textbooks

The textbook hardly needs defining. With the general shortage of mathematics teachers, textbooks are assuming an increasing importance. Whenever mathematicians meet to discuss the modification of mathematical curricula, they invariably stress the urgent need for new and more appropriate textbooks. They see the shortcomings of many of the works at present in use as two-fold: firstly, they do not adequately reflect the significant advances in the subject over the past decades; and secondly, they need to be designed as more complete units, since greater reliance has to be placed on them in the face of the teacher crisis. "The Textbook as a Teaching Aid", *Education Abstracts*, 7, 1955, is the title of a UNESCO contribution to the subject. Pages 1517–24 of the *Encyclopedia of Educational Research*, Macmillan, New York, 3rd edition, 1960, are devoted to textbooks and include a bibliography of 55 items.

There are several services available to educational authorities which may assist in the identification and selection of suitable material. *Current College-Level Book-Selection Service* is the title of a bi-weekly joint publication of the American Library Association and the Council on Library Resources. It reviews books as soon as possible after publication, and draws upon a panel of experts for evaluations. A standard work in the United States is Clapp, Jane, *College Textbooks*, Scarecrow Press, New York, 1960. This is a classified listing of 16,598 textbooks used in 60 colleges and universities in the United States. There are full subject and author indexes, and a directory of publishers and distributors. *Textbooks in Print* is issued annually by the R. R. Bowker Company of New York. It indexes the textbooks of nearly 200 United States publishers. Publishers of this kind of literature have, incidentally, formed an association called the American Textbook Publishers Institute. An example of a specialized listing is the "Guide for the mathematics books of the traveling high school science library of the American Association for the Advancement of Science and the National Science Foundation" by M. F. Willerding, published in *School Science and Mathematics*, **61**, 1961. The bibliography is on pages 101–13. An excellent, an-

H

notated list of works published since 1945 is E. P. Vance's "College text books", *American Mathematical Monthly*, **62**, 265–88, 1955.

In Britain, school-level textbooks are listed in the annual publication, *Education Book Guide* (National Book League, London). This is "designed to bring together the titles of all books suitable for use in schools . . . published in the United Kingdom in a given year". The entries are based on information supplied by the book publishers, a list of which is given in each volume. The section on mathematics is divided into Primary and Secondary.

The Mathematics Section of the Association of Teachers in Colleges and Departments of Education (151 Gower Street, London, W.C.1) has produced a very useful working tool under the title *A List of Books Suitable for Training Colleges*. It is divided into Part 1, "Teaching Method: Primary and Secondary Texts" (1959), and Part 2, "Higher Level Texts, History of Mathematics, etc." (revised 1961). Apart from bibliographical details, including prices, the value of this publication lies in its annotations contributed by people closely associated with mathematics teaching.

Other sources of information on textbooks are the mathematical education periodicals described in Chapter 6. Their review sections serve to keep the reader informed of new publications and their suitability for adoption.

A significant venture in textbook publishing is the Commonwealth and International Library of Science, Technology, Engineering and Liberal Studies (Pergamon Press), of which the present volume forms a part. The very first volume to be published in the Library was a summary of the Southampton Mathematical Conference entitled *On Teaching Mathematics*. The Mathematics Division of the series has already produced a number of textbooks which have been widely adopted.

Pergamon Press is not, of course, alone in this drive to provide much-needed mathematical books. Details of other series will be found in the guides described above.

THE COMMONWEALTH AND INTERNATIONAL LIBRARY OF SCIENCE, TECHNOLOGY, ENGINEERING AND LIBERAL STUDIES

Chairman of the
Honorary Editorial Advisory Board
SIR ROBERT ROBINSON
O.M., F.R.S.

Publisher
ROBERT MAXWELL M.C.

PERGAMON PRESS
OXFORD · LONDON · NEW YORK · PARIS

PERGAMON PRESS LTD.
Headington Hill Hall, Oxford
4 & 5 Fitzroy Square, London W.1

PERGAMON PRESS INC.
122 East 55th Street, New York 22, N.Y.

GAUTHIER-VILLARS
55 Quai des Grands-Augustins, Paris 6

PERGAMON PRESS G.m.b.H.
Kaiserstrasse 75, Frankfurt am Main

Set in Monotype Baskerville by Santype Ltd of Salisbury
and Printed in Great Britain by Tisbury Printing Works Ltd, Salisbury

Aims, Scope and Purpose of the Library

The Commonwealth and International Library of Science, Technology, Engineering and Liberal Studies is designed to provide readers, wherever the English language is used or can be used as a medium of instruction, with a series of low-priced, high-quality, soft-cover textbooks and monographs (each of approximately 128 pages). These will be up to date and written to the highest possible pedagogical and scientific standards, as well as being rapidly and attractively produced and disseminated—with the use of colour printing where appropriate—by employing the most modern printing, binding and mass-distribution techniques.

The books and other teaching aids to be issued by this Library will cover the needs of instructors and pupils in all types of schools and educational establishments (including industry) teaching students on a full and/or part-time basis from the elementary to the most advanced levels.

The books will be published in two styles—a soft-cover edition within the price range of 7s. 6d. to 17s. 6d. ($1.25 to $2.75) and a more expensive edition bound in a hard-cover for library use. The student, the teacher and the instructor will thus be able to acquire, at a moderate price, a personal library in whatever course of study he or she is following.

Books for Industrial Training Schemes to Increase Skills, Productivity and Earnings

To meet the ever-growing and urgent need of manufacturing and business organizations for more skilled workers, technicians, supervisors, and managers in the factory, the office and on the land, the Library will publish, with the help of trade associations, industrial training officers and technical colleges, specially commissioned books suitable for the various training schemes organized by or for industry, commerce and government departments. These books will help readers to increase their skill, efficiency, productivity and earnings.

New Concept in Educational Publishing ; One Thousand Volumes to be Published by 1967 : Speedy Translation and Simultaneous Publication of Suitable Books into Foreign Languages

The Library is a new conception in educational publishing. It will publish original books specially commissioned in a carefully planned series for each subject, giving continuity of study from the introductory stage to the final honours degree standard. Monographs for the post-graduate student and research workers will also be issued, as well as the occasional reprint of an outstanding book, in order to make it available at a low price to the largest possible number of people through our special marketing arrangements.

We shall employ the latest techniques in printing and mass distribution in order to acheive maximum dissemination, sales and income from the books published, including where suitable, their rapid translation and simultaneous publication in French, German, Spanish and Russian through our own or associated publishing houses.

The first 50 volumes of the Library will be issued by the end of 1962; during 1963 a further 150 titles will appear; we expect that by December 1967 a complete Library of over 1000 volumes will have been published. Such a carefully planned, large-scale project in aid of education is unique in the history of the book publishing industry.

New, Modern, Low-priced Textbooks for Students in Great Britain. International Co-operation in Textbook Writing, Publishing and Distribution

In those sciences such as Mathematics, Physics, Chemistry and Biology, which are started at an early age, there will be books suitable for students in Secondary, Grammar and Public Schools in Great Britain covering the work for the new Certificate of Secondary Education and the Ordinary and Advanced Levels of the General Certificate of Education. In all sciences there will be books to meet the examination requirements of the Ordinary National Certificate, Higher National Certificate, City and Guilds and the various other craft and vocational courses, as well as a full range of textbooks required for Diploma and Degree work at Colleges of Technology and Universities.

Similarly books and other teaching aids will also be provided to meet examination requirements in English-speaking countries overseas. Wherever appropriate the textbooks written for British students and courses will be made available to English-speaking students abroad and in particular in countries in the Commonwealth and in the United States.

The help of competent editorial consultants resident in each country will be available to authors at the earliest stages of the drafting of their books to advise them on how their volumes can be made suitable for students in different countries. If, because of differences in curricula and educational practice, substantial changes are needed to make a British textbook suitable in, for example, Australia, or a textbook written by an Indian or an American author suitable for use in the United Kingdom, then the Press will arrange for authors from both countries to collaborate to achieve this.

Co-publishers in the U.S.A. and Canada, The Commonwealth and Other Parts of the World, to Ensure Maximum Possible Dissemination at Low Prices

The Macmillan Company, New York, America's leading educational and general publishers, have already been appointed co-publishers of this Library in the U.S.A. and Canada, and negotiations are also in hand to appoint co-publishers in each of the major Commonwealth countries as well as in Europe, Africa and Asia, to market all books published in the Library exclusively in their country or territory. The co-publishers will assist authors and editors of the Library in the following ways:

(a) By making available to them their editorial contacts, resources and know-how to make the books commissioned for publication in the Library suitable for sale in their country.

(b) By purchasing a substantial quantity of copies of each book for exclusive distribution and sale.

(c) Where useful (in the interest of maximum dissemination), to arrange for or assist with the printing of a special edition, or the entire edition, of a particular textbook.

(d) To use their best endeavours to ensure that the books published in the Library are widely reviewed, publicized, distributed and sold at moderate prices throughout their marketing territory.

International Boards of Eminent Advisory, Consulting and Specialist Editors and Sponsoring Committee of Corporate Members

An Honorary Editorial Advisory Board and a Board of Consulting and Specialist Editors and a Sponsoring Committee under the chairmanship of Sir Robert Robinson, O.M., F.R.S., have been appointed. Some 500 eminent men and women drawn from all walks of life—Universities, Research Institutions, Colleges of Advanced Technology, Industry, Trade Associations, Government Departments, Technical Colleges, Public, Grammar and Secondary Schools, Libraries, and Trade Unions and parliamentarians interested in education—not only from this country but also from abroad—are available to advise by correspondence the editors, the authors, the Press and the national co-publishers to help achieve the high aims and purpose of the Library.

The launching of a library of this magnitude is a bold and exciting adventure. It comes at a time when the thirst for education in all parts of the world is greater than ever. Through education man can get an understanding of his environment and problems and a stimulation of interest which can enrich his life. And, too, if he learns how to apply the results of scientific research, material standards of life can be raised, even in a world of rapidly increasing population embroiled in a great arms race. Some of us hope and believe that through education lies the road to lasting world peace and happiness for all nations and communities, regardless of race, colour or ideology. In the history of education examples can be cited of how one or other famous textbook or author profoundly influenced the education of the period. When at some future time the history of education in the second half of the twentieth century is written, it may well be that the Commonwealth and International Library of Science, Technology, Engineering and Liberal Studies, published by Pergamon Press as a private venture with the co-operation of eminent scientists, educators, industrialists, parliamentarians and others interested in education, will stand out as one of the landmarks.

ROBERT MAXWELL
Publisher at Pergamon Press

Honorary Editorial Advisory Board

8

Exercises

1. What features would you look for in evaluating a bibliography of mathematics?

2. What are the main differences between a library catalogue and a bibliography?

3. If you were trying to sell a bibliography of mathematics to a mathematician, what would be the main points of your sales talk?

CHAPTER 10

Mathematical Books. Part 2: Evaluation and Acquisition

511: Arithmetic
512: Algebra
513–516: Geometry
. 517: Calculus

THE identification of authors and titles of mathematical books was discussed in the previous chapter. The next step is to assess the appropriateness of individual works according to the needs of the enquirer. As there are certain ways in which a preliminary selection can be made without consulting the actual volumes, the question of evaluation will be considered first.

Evaluation

Book Reviews

Some of the bibliographies already described are evaluative in character. That is, by virtue of their arrangement according to user requirements and their annotations they assist making a preliminary selection. A few in fact quote from reviews which have appeared. Book reviews are a useful aid, but great care has to be taken to maintain a sense of proportion when consulting them. It is not unusual to find two reviews at variance in their assessment of the same book. Most reviewers find something to criticize, and one should not therefore always attempt to discover a book which all the reviewers wholeheartedly and unanimously praise. The distinction must also be drawn between reviews proper and publishers' announcements.

Three guides to reviews of mathematical books are:

American Mathematical Society, *Twenty-Volume Index to Mathematical Reviews*, 1940–59, 2 volumes.

Covers the first 20 volumes of *Mathematical Reviews*, and includes cross-references to joint authors. Where there is no personal author the entry is under the editor or title.

Book Review Digest, Wilson, New York, monthly (except February and July).

The entries are arranged in alphabetical order of authors, and include extracts of reviews taken from a wide range of publications. Plus and minus signs are included, when needed and possible, to indicate the verdict of the complete review. Each issue contains a subject and title index, and cumulative volumes are published.

Technical Book Review Index. Special Libraries Association, New York, monthly (Sept.–June).

Quotes reviews from some 1200 scientific, technical and trade journals.

Some Useful Criteria

There is no such thing as the *best* book on a subject. The value of any book to a person seeking information is rigidly determined by its ability to yield that information accurately and in an understandable manner. So much depends on the user's own ability to interpret the information presented in books that it is virtually impossible to name a book "the best". Certain criteria do nonetheless exist for the evaluation of books in general terms. If applied as a matter of habit when handling books they can frequently save frustration and minimize the risk of acquiring erroneous data.

The date of publication is probably the first thing to look for. Here it is necessary to know the difference between a new edition, a reprint and a re-impression. A new edition really constitutes a new book, earlier editions having been superseded by the inclusion of new material, corrections and re-shaping. A re-impression, on the other hand, is exactly the same as the work previously published. It is in fact printed from the same type. A reprint is produced from new type and contains very few, if any, textual amendments. Unfortunately, many books are called new editions by their publishers when only the slightest textual differences exist. A detailed comparison of the actual volumes is the only sure test. Some indication of the qualifications of the author are frequently given either on the title-page of the book or on its paper jacket. Depending on the type of book will be the desirability of a subject and/or author index. The value of many potentially useful books is seriously reduced by the lack of an adequate index. The inclusion of a bibliography or guide to further reading is another desirable feature, though this again depends on the nature of the individual work. Publication dates of the items comprising such a bibliography can also serve as a guide to the currency of the textual matter.

Acquisition

Books in Print

Acquisition of books will naturally be influenced to some extent by their cost. Whilst library services are constantly improving, it is not, however, their intention to obviate the need for personal collections. Efforts are being made to encourage people to buy their own copies. Special paperback editions are produced at reasonable prices, and several good mathematical series are available in such editions. There is even a catalogue of paperbacks in print. Most professional societies offer specially reduced rates for their publications to members. Unfortunately, this question of cost often adversely influences the selection of books for purchase, though it is emphatically *not* true to say that cheaper books are necessarily inferior.

Guides are published to books which are currently available. *Books in Print: an Author–Title–Series Index to the Publishers' Trade List Annual* is issued by Bowker, New York. For those using a subject approach, Bowker also publish a *Subject Guide to Books in Print*, which again is an index to their *Publishers' Trade List Annual*. Their new *American Scientific Books 1960–1962*, edited by Phyllis B. Steckler, is a cumulation from the *American Book Publishing Record* and is arranged by the Dewey classification. Author and title indexes are included.

Covering the British book trade is *The Reference Catalogue of Current Literature*, Whitaker, London. The 1961 edition is in 2 volumes: 1, *Author Index*, and 2, *Title Index*. It contains 350,000 entries, and thus constitutes a comprehensive index of books in print and on sale in the United Kingdom. The list can be updated by reference to the quarterly and annual issues of *Whitaker's Cumulative Book List*, and the weekly *Bookseller*, also published by Whitaker. The *British National Bibliography* is described on pp. 103–104.

After drawing up a list of titles selected from mathematical bibliographies, therefore, one can easily discover which of them may be purchased without difficulty.

Great care has to be exercised in purchasing second-hand books, particularly mathematical textbooks. Existing works are rapidly being superseded by those which match the revised curricula. Many older mathematical works still do have current value, but as far as textbooks are concerned early volumes are now usually only of academic interest.

Out-of-Print Books

Obtaining out-of-print works can be a protracted affair, and the enquirer is advised to put the matter in the hands of a good bookseller. There does exist a unique catalogue of older mathematical texts. This is the *Bibliotheca Chemico-Mathematica: Catalogue of Works in Many Tongues on Exact and Applied Science, with a Subject-Index*, compiled and annotated by H. Z. (Heinrich Zeitlinger) and H. C. S. (Henry Cecil Sotheran), 2 volumes, Henry Sotheran and Co., London, 1921. Entries are arranged in an author sequence. The preface states that "As the Catalogue is one of actual books for sale it is of course not complete, but it is believed that few of the great books will be found lacking". The *First Supplement*, published in 1932, has a classified arrangement with mathematics occupying 45 pages. The *Second Supplement* in two volumes was published in 1937, and comprises 22,895 items. Again the arrangement is classified, with mathematics this time having 133 pages. The annotations are very valuable and the entire catalogue is a work of great scholarship. The *Mathematical Gazette* is quoted as saying that "Sotheran's lists are almost a liberal education in mathematical history". Separate mathematics lists contain additional titles.

Keeping Abreast of New Mathematical Books
Current Bibliographies

Mathematical Reviews is of course the main guide to new publications in all branches of mathematics, and from all countries. The American Mathematical Society also produces a list of *New Publications* as a separate item which is available from the Society's Special Projects Department. It was previously published as a regular feature of the *Bulletin*. The Society's own new publications are still given in the *Bulletin*. In 1961 a new service was commenced by the American Bibliographic Service, Darien, Connecticut. It is a *Quarterly Checklist of Mathematica: an International Index of Current Books, Monographs, Brochures and Separates*.

As a matter of course, the mathematician will read the book reviews in the periodicals which he has selected to peruse. Sometimes overlooked are the lists of new publications which the editor has received for reviewing purposes. Space limitations alone will prevent the review of all the material he receives, and whilst he will attempt to cover the books of widest interest, he is bound to pass over many others. The more specialized the nature of the periodical, the more complete the reviewing coverage is likely to be for the field in question.

Most of the larger libraries, including those of industrial organizations, issue their own accessions lists, often in a classified arrangement. In using them, care should be taken to consult all possible headings under which relevant titles could appear. Titles on linear programming, for example, might be placed under the management heading, or those on econometrics under economics. Every subject specialist should consider it a professional responsibility to acquaint the head of the library service which he regularly uses with details of publications he feels should be purchased. Book selection policies, though always subject to financial considerations, are fundamentally geared to the known and potential requirements of library users. Clearly, the best way in which a librarian can become aware of his clients' needs is from the clients themselves. When a librarian is not sure from the bare bibliographical details of a book whether or not it is worthy of inclusion in his stock he may order an examination copy from the publisher. Close liaison with the librarian in this connection will prove to be of mutual benefit.

There are a number of comprehensive bibliographies, issued on a continuing basis, which are used by librarians as book selection tools, and which are therefore readily accessible to the mathematician.

The *Cumulative Book Index* (C.B.I.) is issued monthly by the H. W. Wilson Company, New York, and later appears in cumulated volumes designed to reduce the work of consultation. The entries are arranged in one dictionary-type sequence of authors, titles and subjects. Cross-references between subject headings are provided. On looking up geometry, for example, one finds the following headings used:

    Geometry
        see also
      Curves
      Topology
      Trigonometry
    Geometry, Algebraic
        see also
      Geometry, Analytic
      Topology
    Geometry, Analytic
        see also
      Conic sections
      Curves
      Geometry, Algebraic
    Geometry, Descriptive
        see also
      Perspective
    Etc.

The *British National Bibliography* (B.N.B.) is arranged according to the Dewey Decimal Classification, and is issued weekly with subsequent cumulations. Whilst the C.B.I. covers all English-language publications, the entries in the B.N.B. are based on copies received at the British Museum Copyright Office. The publication comprises two sections. In the first the entries are arranged according to the Dewey Decimal Classification. The second section contains entries and references under authors, titles, editors, series and subjects in one alphabetical sequence. First issues of new periodicals and those which have changed their titles are also noted. *British Book News* is largely devoted to reviews of books and new periodicals. Every review bears a Dewey class number, and entries are arranged in a classified order.

Since 1956 the Library of Congress in Washington, D.C., has published the *National Union Catalog: a Cumulative Author List Representing Library of Congress Printed Cards and Titles Reported by other American Libraries*. This extremely valuable work reporting the current intake of the largest libraries in the United States is published monthly with quarterly and annual cumulations. It was previously known as the *Library of Congress Catalog —Books: Authors*, and constitutes an updating service to the multi-volume printed catalogue of the Library of Congress. Its quarterly counterpart which began publication in 1950 is the *Library of Congress Catalog—Books: Subjects*. This is arranged in an alphabetical sequence of subject headings and a location in at least one library in the United States is given for each title, additional locations being supplied in the *National Union Catalog*. An idea of the type of headings used can be gained from examining what one finds on looking up algebra in a sample issue:

ALGEBRA
    —Problems, Exercises, etc.
    —Study and Teaching
ALGEBRA, ABSTRACT
ALGEBRA, UNIVERSAL
ALGEBRA OF LOGIC see Logic, symbolic and mathematical
ALGEBRAIC CONFIGURATIONS IN HYPERSPACE
    see Hyperspace
ALGEBRAIC FIELDS see Fields, Algebraic
ALGEBRAIC FUNCTIONS see Functions, Algebraic

ALGEBRAIC TOPOLOGY
ALGEBRAS, LIE see Lie algebras

Of less impressive proportions, but nonetheless very useful, is the *Aslib Book List: a Monthly List of Recommended Scientific and Technical Books with Annotations*, Aslib, London. The sub-title is an accurate description, for brief notes on each work amplify the grading which has been made by the reviewing subject special-list, according to the following plan:

A: Books suitable for general readers; treating their subject in an introductory, elementary, or general manner.
B: Books of intermediate technical standard or students' textbooks.
C: Books of an advanced or highly technical character.
D: Directories, dictionaries, handbooks, lists and catalogues, encyclo-pedias, yearbooks, and similar publications.

The entries are arranged by the Universal Decimal Classification.

National bibliographies are available for most of the major countries. They are particularly useful when their entries are grouped by subject, as in the German (F.G.R.) *Deutsche Biblio-graphie*. *Biblio* covers all books published in French, and its arrangement resembles that of the *Cumulative Book Index*. *Bibliographical Services Throughout the World* (see p. 17) provides information on other such services.

Publishers' Announcements

Any publisher will be pleased to keep a potential customer informed of his new mathematical books by mailing descriptive sheets and subject lists. All that is necessary is to notify the publisher of one's particular interests. Selection of publishers should be carefully made so as to avoid unnecessary wastage on both sides. This can be done by seeing who has published existing works on one's subject. References in the present volume include the names of most of the major mathematical publishers. Their addresses may be obtained from directories such as the *American Booktrade Directory* (Bowker, New York), and the *Publishers' International Year Book*, *World Directory of Book Publishers* (Wales, London).

Exercises

1. How would you evaluate a new mathematical book which you are personally able to examine?
2. Describe two bibliographical services which may be used for checking whether or not particular books are in print.
3. What steps would you take to ensure that you were kept informed of new mathematical books published in the English language?

CHAPTER 11

Probability and Statistics

519: Probabilities and mathematical statistics

Introduction

The recent National Science Foundation report on *Employment in Professional Mathematical Work in Industry and Government*, having briefly defined statistics as "a science based on the mathematical theory of probability", goes on to state that: "In recent years there has been a growth in the application of mathematical statistics to physical and biological phenomena and to business management problems by the use of statistical theory in the design of experiments, in quality control, and in evaluating the likely result of proposed plans of action." This activity is reflected in the vast amount of relevant literature which continues to be published. Both this chapter and the one which follows provide a guide to sources of information in the field.

Current trends are reported in the *Proceedings of the Berkeley Symposia on Mathematical Statistics and Probability*, published by the University of California Press. The Berkeley Symposia have been held every 5 years since 1945. They have now become fully international, and the *Proceedings* of the fourth symposium, held in 1960, comprise 5 large volumes covering the theory of statistics; probability theory; and applications in astronomy, physics, biology, health, econometrics, industrial research, and psychometry. Contributions are followed by lists of references.

A series of "Studies in the history of probability and statistics" in *Biometrika* was started by F. N. David in Volume 42.

Before proceeding to an examination of the various sources of information, attention should be drawn to the Dewey 310 class. This caters for statistics under the social sciences heading, and is divided into 311 Statistical Method; 312 Demography; 313–319 General Statistics of Specific Countries.

Books and Bibliographies

Maurice G. Kendall is a name which recurs in the literature of statistics. Kendall has published many important contributions to the subject individually, and has also combined with Alison G. Doig in the joint production of *A Bibliography of Statistical Literature*, a work designed to cover the literature of probability and statistics from the 16th Century to 1958. Volume 1 (Oliver and Boyd, Edinburgh, 1962) deals with the period 1950–58. Volume 2 takes 1940–49, leaving the remaining years for Volume 3. The references—some 30,000 in the complete work—are arranged alphabetically by author.

As part of its programme of education the International Statistical Institute undertakes bibliographical work. Its *Bibliography of Basic Texts and Monographs on Statistical Methods*, The Hague, 1951, is a result of this policy. It comprises about 100 English-language items dealing both with methods and their application. In each case, bibliographical details are followed by a list of chapter headings and extracts from reviews which have appeared in the *Journal of the Royal Statistical Society*, the *Journal of the American Statistical Association*, *Biometrika*, and *Sankhyā*.

Other useful bibliographies in the field are:

BUROS, OSCAR KRISEN (ed.), *Statistical Methodology Reviews 1941–1950.* Wiley, New York, 1951.

Quotes reviews of 342 books, written in English and published or reviewed in the period, dealing with statistical methods "and such closely related subjects as probability and mathematics of statistics". Earlier years are covered by *Research and Statistical Methodology Books and Reviews, 1933–38* and *The Second Yearbook of Research and Statistical Methodology, Books and Reviews*, published by Gryphon, Highland Park, New Jersey, in 1938 and 1941 respectively.

DEMING, LOLA S., "Selected bibliography of statistical literature, 1930–1957", *Journal of Research of the National Bureau of Standards—B: Mathematics and Mathematical Physics.*
This is a series of bibliographies of which the first, on correlation and regression theory, appeared on pages 55–68 of the Jan.–Mar., 1960 issue. The entries are based on the N.B.S. Statistical Engineering Laboratory's card file of abstracts compiled from the *Zentralblatt für Mathematik* (for the years 1930–39) and *Mathematical Reviews* (from 1940 onward).

SAVAGE, I. RICHARD, *Bibliography of Nonparametric Statistics*, Harvard University Press, Cambridge, Mass., 1962.
This revised edition contains about 3000 entries published up to April, 1961.

A statistical bibliography of a different kind, published by Cambridge University Press, is *A Bibliography of the Statistical and Other Writings of Karl Pearson*, compiled by G. M. Morant and B. L. Welch. The same publisher has also issued *Karl Pearson: An Appreciation of Some Aspects of his Life and Work* by E. S. Pearson.

As far as textbooks and monographs are concerned, no more can be done here than to indicate a few important new or recently revised works. The International Statistical Institute's *Bibliography of Basic Texts* contains details of many of the well-tried books. Textbooks dealing with the various applications of statistics are mentioned in their appropriate sections in both this chapter and the next.

A useful and reasonably-priced work on elementary statistics suitable for early university work is *Elementary Statistical Exercises*, by F. N. David and E. S. Pearson, Cambridge University Press, 1961. Other titles are *Statistics: an Introduction* by D. A. S. Fraser, Wiley, 1959; *An Introduction to Mathematical Statistics* by H. D. Brunk, Ginn, 1960; *Introduction to Mathematical Statistics* by P. Hoel, Wiley, 2nd edition, 1954; *Elements of Mathematical Statistics* by J. F. Ractliffe, Oxford University Press, London, 1962; and *A First Course in Mathematical Statistics* by C. E. Weatherburn, Cambridge University Press, London, 1961. Two books by Samuel S. Wilks should also be noted:

Elementary Statistical Analysis (Princeton University Press, 1949), and *Mathematical Statistics* (Wiley, New York, 1962).

Of particular importance is *The Advanced Theory of Statistics* by M. G. Kendall and Alan Stuart, Griffin, London. The complete work will be in 3 volumes, of which 2 have so far appeared; 1, *Distribution Theory*, 1958; and 2, *Statistical Inference and Statistical Relationship*, revised edition, 1961. Volume 3, dealing with statistical planning and analysis, and time series, is scheduled to appear in 1964.

Good introductions to probability theory which may be listed together are:

> FELLER, WILLIAM, *An Introduction to Probability Theory and its Applications*, Wiley, New York, 2nd edition, 1957.
>
> GNEDENKO, B. V. and KHINCHIN, A. YA, *An Elementary Introduction to the Theory of Probability*. Freeman, San Francisco and London, 1961.
>
> KOLMOGOROV, A. N., *Foundations of Probability Theory*. Chelsea, New York, 1950.
>
> PARZEN, EMANUEL, *Modern Probability Theory and its Applications*. Wiley, New York, 1960.
> (Parzen's *Stochastic Processes* was published by Holden-Day, San Francisco, in 1962.)
>
> TUCKER, HOWARD G., *An Introduction to Probability and Mathematical Statistics*. Academic Press, New York, 1962.

Probability Theory by M. Loeve, Van Nostrand, Princeton, New Jersey, 2nd edition, 1960, covers more advanced work.

Dictionaries

The year 1960 saw the publication of two valuable reference works for statisticians. The second edition of *A Dictionary of Statistical Terms* by Maurice G. Kendall and William R. Buckland was published for the International Statistical Institute by Oliver and Boyd, Edinburgh. Following the definitions in the first edition, 1957, were four glossaries of equivalent terms arranged in alphabetical order for each of the four other working languages of the Institute: French, German, Italian and Spanish. The new work has in addition a *combined* glossary in which the English terms determine the alphabetical order. This new revised, combined glossary is also available separately. In the same year the Hungarian Central Statistical Office published its *Statistical Dictionary* containing some 1700 terms in 7 languages. The volume consists of two parts; 1, basic tables arranged in columns, with the Russian terms determining the alphabetical (Cyrillic) sequence, and 2, alphabetical indices of the individual languages—Bulgarian, Czech, English, German, Hungarian, Polish and Russian. Entries in an appended bibliography include the *Multilingual Demographic Dictionary*, United Nations, New York, 1958; *Statistical Vocabulary*, Inter American Statistical Institute, Washington, 1950; and V. Trapp's *Statistisches Wörterbuch, Englisch-Deutsch* (*Einzelschriften der Deutschen Statistischen Gesellschaft*, Heft No. 2, Munich, 1956).

Abstracts and Periodicals

Mathematical statistics is well covered by scientific periodicals, most of which are published by professional societies. More recently inaugurated titles represent those fields of application in which there has been particularly rapid development in recent years. A good example is *Technometrics*.

Statisticians are fortunate in having the bulk of this literature abstracted for them in the *International Journal of Abstracts: Statistical Theory and Method* (a journal of the International Statistical Institute), Oliver and Boyd, Edinburgh, quarterly. Its expressed aim is to give complete coverage of published papers in the field of statistical theory and newly published contributions to statistical method. All articles in the following journals, which are wholly devoted to the field, are abstracted: *Annals of Mathematical Statistics* (Institute of Mathematical Statistics, Stanford University), *Biometrika*, *Journal of the Royal Statistical Society* (Series B), *Bulletin of Mathematical Statistics*, and *Annals of the Institute of Statistical Mathematics*. Selective abstracting of other journals is also undertaken: *Biometrics*, *Metrika*, *Metron*, *International Statistical Institute Review*, *Technometrics*, and *Sankhyā*. The abstracts, which are all in English irrespective of the language

J

of the original, are about 400 words long. Addresses of authors are given so as to facilitate communications.

Individual statistical journals do, of course, include review sections of their own. The Royal Statistical Society's *Applied Statistics* (Oliver and Boyd) and the *Journal of the American Statistical Association* are examples.

Translations

Foreign works, particularly Russian, on probability and statistics are currently receiving attention commensurate with that already described for mathematics in general.

The Society for Industrial and Applied Mathematics, for example, is producing a translation of the Russian journal *Teoriya Veroyatnosti i ee Primeneniye* under the title *Theory of Probability and its Applications*. It contains papers on the theory and application of probability, statistics, and stochastic processes, and is published through a grant-in-aid by the National Science Foundation.

Selected Translation in Mathematical Statistics and Probability also benefits from an N.S.F. grant. Volume 1 (1961) contains 25 papers, and Volume 2 (1962) has 19 papers. The translations are published for the Institute of Mathematical Statistics by the American Mathematical Society.

Statistical Tables

An important contribution to this subject, and one which should be in the personal library of every statistician, is Greenwood, J. Arthur and Hartley, H. O. *Guide to Tables in Mathematical Statistics*. Princeton University Press, 1962. It is a catalogue of tables of which the majority were published between 1900 and 1954. Full references are provided for each entry, and the work is well equipped with detailed author and subject indexes. A useful feature is an appendix which contains the contents lists of a number of books of tables. This check list is of such practical value, as well as being a handy bibliography, that the works represented are given here:

ARKIN, H. and COLTON, R. R., *Tables for Statisticians*
BURINGTON, R. S. and MAY, D. C., *Probability and Statistics*
CZECHOWSKI, T. *et al.*, *Tablice Statystyczne*
DIXON, W. J. and MASSEY, F. J., *Introduction to Statistical Analysis*
FISHER, R. A. and YATES, F., *Statistical Tables*
GLOVER, J. W., *Tables of Applied Mathematics*
GRAF, U. and HENNING, H. J., *Formeln und Tabellen*
HALD, A., *Statistical Tables and Formulas*
JAHNKE, P. R. E. and EMDE, F., *Funktionentafeln—Tables of Functions*
KELLEY, T. L., *The Kelley Statistical Tables*
KITAGAWA, T. and MITOME, M., *Tables for the Design of Factorial Experiments*
LINDLEY, D. V. and MILLER, J. C. P., *Cambridge Elementary Statistical Tables*
PEARSON, E. S. and HARTLEY, H. O., *Biometrika Tables for Statisticians*
PEARSON, K., *Tables for S. & B. I & II*
SIEGEL, S., *Nonparametric Statistics*
VIANELLI, S., *Prontuari per Calcoli Statistici*

An important collection which was published after the compilation of this Guide is D. B. Owen's *Handbook of Statistical Tables*, Addison–Wesley, Reading, Mass., 1962.

Tables are frequently published for the first time in scientific journals. It is, however, worth noting here that copies of such tables can sometimes be purchased separately. An example is the series of *New Statistical Tables: Separates Re-issued from Biometrika* which can be obtained from the Biometrika Office, University College, London.

Societies

The International Statistical Institute (The Hague, Holland) was founded in 1855. Its aims are the improvement of statistical methods, to which end it arranges conferences, assists with educational programmes and publishes several periodicals, including a *Bulletin* and *Review*. Details of its other titles are given elsewhere in this chapter. Further information will be found in J. W. Nixon's *A History of the International Statistical Institute*, 1885–1960 (The Institute, Hague, 1960), which supplements F. Zahn's *50 Années de l'Institut International de Statistique* (1934).

National societies exist in most industrial countries of the world.

As a minimum activity they publish contributions to the subject, and are able to supply current information on all aspects of statistics in their respective countries.

Important American bodies are:
American Statistical Association,
810 18th Street, N.W., Washington 6, D.C.

Institute of Mathematical Statistics,
Department of Statistics, University of North Carolina.

Inter American Statistical Institute,
Pan American Union, Washington 6, D.C.

In Britain, the Royal Statistical Society, 21 Bentinck Street, London, W.1, was founded in 1834. Its *Journal* and *Applied Statistics* are important publications. The Society possesses a fine library. The Yule Library, based on books bequeathed by George Udny Yule, includes rare and early statistical publications. Among Yule's writings, his *Statistical Study of Literary Vocabulary* (1944) may be noted. In it the author discusses the application of statistical methods to the study of vocabulary in cases of disputep authorship.

Applications

The applications of statistical techniques are very numerous and have rightly occasioned a substantial amount of literature. As far as the organization of this information is concerned, its distribution in the schedules of the Dewey Decimal Classification is dependent upon the way the material is treated. Where the emphasis is on the subject to which statistics is applied, it may be classified with that subject. The 16th edition of Dewey, for example, allocates the notation 330.18, in the Economics class, to Econometrics. Similarly, Quality Control in Production is classified in the Production Management class at 658.562. On the other hand, such books as R. A. Fisher's *Statistical Methods for Research Workers*, although emphasizing statistical methods in biology, are so fundamental as to be more properly classified in the 519 class. The same could also be said for R. G. D. Steel and J. H. Torrie's *Principles and Procedures of Statistics*.

Provided these considerations are borne in mind it is appropriate to describe some of the sources of information on applied statistics here. They may be conveniently grouped under three broad headings, but it should be stressed that the following is not intended as a complete survey. The object is rather to indicate the type of material which is available.

Biology, Agriculture, Medicine

Biometrics is the application of statistical methods to the field of biological research, and is a subject which has seen a rapid growth in recent years. Undoubtedly one of the best books on statistical methods in biology is Sir Ronald Aylmer Fisher's *Statistical Methods for Research Workers*, published by Oliver and Boyd. The latest edition should always be sought. It includes a bibliography. Another classic by the same author is *The Design of Experiments*, Oliver and Boyd, Edinburgh, 7th edition, 1960; and a third book of the same high calibre is *Statistical Methods and Scientific Inference*, Oliver and Boyd, Edinburgh, 2nd edition. 1959. *Biomathematics: the Principles of Mathematics for Students of Biological Science* by Cedric Austen Bardell Smith, Griffin, London, 1954, is a standard reference work suitable for university courses. Also noteworthy are K. Mather's *Statistical Analysis in Biology*, Methuen, London, 2nd edition, 1946, and Huldah Bancroft's *Introduction to Biostatistics*, Hoeber (Harrap), New York, 1957.

A special application is in the study of genetics. The proceedings of an international symposium, held at Ottawa in 1958 and jointly sponsored by the Biometrics Society and the International Union of Biological Sciences, were edited by Oscar Kempthorne and published in 1960 under the title *Biometrical Genetics*. Kempthorne's *An Introduction to Genetic Statistics* was published by Wiley, New York, in 1957. A more recent publication, which contains a good bibliography, is Bailey, Norman T. J., *Introduction to the Mathematical Theory of Genetic Linkage*, Oxford University Press, 1961. Bailey's *Statistical Methods in Biology*, English

Universities Press, London, 1959, is a good introductory textbook. Two more works specifically on genetics are Falconer, D. S., *Introduction to Quantitative Genetics*, Oliver and Boyd, Edinburgh, 1960, and Moran, P. A. P., *The Statistical Processes of Evolutionary Theory*, Oxford University Press, 1962.

Covering another specialized field is *An Annotated Bibliography on the Uses of Statistics in Ecology: a Search of 31 Periodicals* by Vincent Schultz, U.S. Atomic Energy Commission, Washington, D.C., 1961.

Agricultural applications are dealt with in D. J. Finney's *An Introduction to Statistical Science in Agriculture*, Oliver and Boyd, Edinburgh, 2nd edition, 1962. Other useful textbooks include Sampford, M. R., *An Introduction to Sampling Theory with Applications to Agriculture*, also published by Oliver and Boyd in 1962, and *Statistical Methods for Agricultural Workers* by V. G. Panse and P. V. Sukhatme, published by the Indian Council of Agricultural Research in 1955.

Tables primarily used in the above applications have been collected together by R. A. Fisher and F. Yates in *Statistical Tables for Biological, Agricultural and Medical Research*, Oliver and Boyd, of which several editions have appeared.

The importance of agricultural statistics is recognized at government level. In Britain, for instance, the Agricultural Research Council controls the Statistics Service of Cambridge University's School of Agriculture. Similar functions are performed in the United States by the Agricultural Research Service of the U.S. Department of Agriculture.

Reviewing Austin Bradford Hill's *Principles of Medical Statistics* (*The Lancet*, London, 7th edition, 1961), in the *Journal of the Royal Statistical Society*, P. D. Oldham describes it as "unquestionably the best source of statistical knowledge for those working in the field of medicine".

Economics and Business

According to Lange (see below), econometrics "tries by mathematical and statistical methods to give concrete quantitative expression to the general schematic laws established by economic theory". Further elaboration of its nature and functions will be found in two recent articles in *Applied Statistics*. The first, by Eric Shankleman, is entitled "What is econometrics?" (**3**, 85–89, 1954), and the second "The scope and limitations of econometrics", by L. R. Klein (**6**, 1–17, 1957). Although Klein wishes to rule out pure mathematical economics as being simply economic theory in a particular form, there is one recent work on the subject which may be noted in passing. It is Reghinos Theocharis' *Early Developments in Mathematical Economics*, Macmillan, London, 1961. This covers the history up to the mid-19th Century work of the famous A. A. Cournot, and includes a bibliography. Also worthy of note is the second edition of R. G. D. Allen's *Mathematical Economics*, which was published by Macmillan in 1959.

Klein's paper, which considers econometrics as the theory and application of measurement in economics guided by an underlying mathematically expressed model, is followed by a useful selective bibliography in which the entries are grouped under headings: general, demand analysis, production and cost functions, export–import functions, aggregative models, sample survey information, and miscellaneous. The following are all useful works on the subject:

BEACH, E. F., *Economic Models: an Exposition*. Wiley, New York, 1957. Includes many references to further readings.

KLEIN, L. R., *Econometrics*. New York, 1955.

LANGE, OSKAR, *Introduction to Econometrics*. London, 1959. Translated from the Polish.

TINBERGEN, JAN, *Introduction to Econometrics*. London, 1953.

TINBERGEN, J. and Bos, H. C., *Mathematical Models of Economic Growth*. New York, 1962.

Econometrica, published quarterly by the Econometric Society (Box 1262, Yale University, New Haven, Conn.), includes a regular book review feature. A list of the Society's members is given in the October 1961 issue.

A good textbook for beginners in business applications is Freund, John E. and Williams, Frank J., *Modern Business Statistics*. Pitman, London, 1959. Freund has also authored *Mathematical Statistics*, published by Prentice-Hall in 1962.

Industry

Technometrics is the title of a quarterly journal of statistics for the physical, chemical, and engineering sciences. It was jointly launched by the American Statistical Association and the American Society for Quality Control, in 1959. In the following year, Academic Press commenced publication of the *Journal of Mathematical Analysis and Applications* which covers the mathematical treatment of questions arising in physics, chemistry, biology and engineering.

Before further mention is made of statistical quality control, it must be noted that the 16th edition of the Dewey Decimal Classification has allocated the notation 658.562 in the Production Management class to Quality Control in Production. The enquirer should, therefore, not fail to examine this section of a library's stock.

The Industrial Statistics Committee of the Eastman Kodak Company has produced a volume of *Symbols, Definitions and Tables for Industrial Statistics and Quality Control* (Institute of Technology, Rochester, New York). Another useful reference work is Leslie E. Simon's *An Engineer's Manual of Statistical Methods*, Wiley, New York, repr. 1950, which deals especially with statistical methodology and industrial quality control. Rapid progress has been made in the field of statistical quality control and the application of sampling theory to the question of life expectancy of products, and several good textbooks such as Acheson J. Duncan's *Quality Control and Industrial Statistics* are available.

Among more general works are Kenneth A. Brownlee's *Statistical Theory and Methodology in Science and Engineering*,

A. Hald's *Statistical Theory with Engineering Applications*, and Paradine and Rivett's *Statistics for Technologists*.

A compact and inexpensive *Chemist's Introduction to Statistics, Theory of Error and Design of Experiments* by D. A. Pantony was published by the Royal Institute of Chemistry, London, in 1961.

Other Applications

It would be possible to fill many more pages with information on yet more applications of statistics. As the scientific validity of statistical techniques becomes more generally accepted, their circle of influence becomes more widespread. Psychology and education, for example, can claim such works as A. L. Edwards' *Statistical Analysis for Students of Psychology and Education*, J. P. Guilford's *Fundamental Statistics in Psychology and Education*, Quinn McNemar's *Psychological Statistics*, and William S. Ray's *Statistics in Psychological Research*.

As new applications are elaborated they are reported in the periodical press, and those who wish to keep abreast must make a habit of perusing the current issues of some of the titles listed earlier in the chapter.

Exercises

1. Explain why not all books on the application of statistics appear in class 519 of the Dewey Decimal Classification. Illustrate your answer with examples.
2. Describe any *three* of the following:
   (a) a statistical dictionary
   (b) a statistical bibliography
   (c) a statistical society
   (d) a statistical abstracting publication
   (e) a guide to statistical tables.
3. What sources would you use in compiling a bibliography of statistical quality control?

CHAPTER 12

Operational Research and Related Techniques

Dewey 519.9 Class

Operational Research

Introduction

Operational research is the term used to describe the application of scientific principles to business problems. In the United States it is more commonly referred to as *operations research*. Perhaps it is in order to overcome this diversity of teminology that the initials O.R. have been commonly adopted. Of the many definitions of O.R. that have been proposed, the following, contained in a brochure issued by the Operations Research Society of America, is as helpful as any:

> *Operations research is the science that is devoted to describing, under-standing, and predicting the behavior of . . . man-machine systems operating in natural environments.*

Management is often defined as a decision-making process. The manager's function is the efficient blending of men, materials and money. So many variable factors are involved in the succession of alternative lines of action which confront him that in making his choice he is driven to rely on his intuition and what is commonly called *business acumen*. He is compelled to follow the road he knows to be safe. Whilst safe, it may not necessarily be the most efficient and profitable. However, to grasp the implications of all the alternatives would be about as difficult as fixing in his mind the pattern produced by a constantly changing kaleidoscope. Such problems are found, for example, in the allocation of resources, in inventory management, and in the scheduling of production and shipments. It is in these and similar fields that O.R. has come to the aid of management by applying systematic quantitative analysis to the decision-making process. All the many factors involved in each set of alternatives are reduced by statistical techniques to a scientifically organized array of data. From these data a mathematical model may be produced and subjected to tests and analysis in order to establish a reliable basis upon which decisions may be founded.

This, of course, is only a very brief outline of the essential nature of O.R. Its breadth is extending rapidly, and its implications are as wide as those of management itself. They range from military strategy to coal mining, from hospital administration to the steel industry.

Introductions to the subject are indicated later, in the section on books. A concise background periodical article which is well worth consulting is "Mathematics for decision makers" by R. K. Gaumnitz and O. H. Brownlee. It appeared on pages 48–56 of the May–June 1956 issue of *Harvard Business Review*. An account of the early history of O.R. is given in McCloskey and Trefethen's *Operations Research for Management* (see p. 124).

Societies

The International Federation of Operational Research Societies was founded in 1959 through the co-operation of the national societies of France, the United Kingdom, and the United States. A further seven members were accepted in 1960, and by March 1962 fourteen countries were represented, namely:

Argentina:	Sociedad Argentina de Investigación Operativa (Buenos Aires)
Australia:	Australian Joint Council for Operational Research (Sydney)
Belgium:	Société Belge pour l'Application des Méthodes Scientifiques de Gestion (Brussels)
Canada:	Canadian Operational Research Society (Ottawa)
France:	Société Française de Recherche Operationnelle (Paris)
Germany:	Deutsche Gesellschaft für Unternehmensforschung (Bonn)
Holland:	Sectie Operationele Research, of the Dutch Statistical Society (Vinklaan 1, Son)

India:	Operational Research Society of India (Delhi)
Italy:	Associazione Italiana di Ricerca Operativa (Rome)
Japan:	Operations Research Society of Japan (Tokyo)
Norway:	Norwegian Operations Research Society (Oslo)
Sweden:	Swedish Operations Research Association (Stockholm)
U.K.:	Operational Research Society Ltd. (London)
U.S.A.:	Operations Research Society of America (Baltimore, Maryland)

Total membership of the Federation is well over 5000 individuals.

Every three years an International Conference on Operational Research is held. The first was held in Oxford, England in 1957, and succeeding ones in Aix-en-Provence, and Oslo. Conference *Proceedings* are published and form an important contribution to the subject. The *Proceedings of the Second International Conference on Operational Research*, edited by J. Banbury and J. Maifland, were published by Wiley in 1961.

From the outset the need for an effective medium for disseminating O.R. information was recognized. The very first *Annual Report* of I.F.O.R.S. states that: "Preliminary consideration has been given to methods of compiling and distributing international abstracts". The first issue of *International Abstracts in Operations Research* appeared in 1961. In order to provide world-wide coverage, reviewers in the member societies of the Federation analyse the literature of their respective countries, and additional countries are also represented. The abstracts are written in English, and there are comprehensive subject and author indexes. The subject indexes are based on "index titles" prepared by the abstractors.

The Operations Research Society of America (Mount Royal and Guilford Avenues, Baltimore 2, Maryland) was founded in 1952, and in its first 10 years attracted a membership of over 3500. It now has several local sections. Normally, two national meetings are held annually in different cities of the United States and Canada, as well as sectional meetings. The Society is responsible each year for awarding the $1000 Lanchester Prize for the best English-language paper in O.R. In addition to the periodicals included in the list below a *Directory* of members is published annually, and the Society also sponsors the publication of a series of books called *Publications in Operations Research*.

The Operational Research Society, Ltd. (64 Cannon Street, London, E.C.4, England) developed from a club founded in 1947. It now has over 650 members whose common interest is "the development and extension of operational research as a branch of science". The activities of the Society include the publication of *Operational Research Quarterly* (Pergamon Press), a comprehensive library, an information service dealing with all aspects of O.R., technical meetings, and assistance in the setting up of courses in O.R. at universities and colleges of technology.

Bibliographies

Operational Research, being a recently developed science, has been able to profit from the experience of the older sciences in tackling the problem of communication. As has already been noted, provision was made soon after the foundation of the International Federation for the publication of an abstracting journal of world-wide scope. There still remained, however, a need for retrospective indexing, and the following two publications fill in a large part of the gap:

> BATCHELOR, JAMES H., *Operations Research: An Annotated Bibliography*. St. Louis University Press, 2nd edition, 1959.
> CASE INSTITUTE, OPERATIONS RESEARCH GROUP. *A Comprehensive Bibliography on Operations Research*. Wiley, New York, 1958.

Apart from the *International Abstracts*, information on current literature is given in many of the periodicals which cover the subject and in some of the abstracting and indexing publications described in Chapter 4. An annual review of progress in the field is also published:

> *Progress in Operations Research*, Russell L. Ackoff (ed.), Wiley, New York, Volume 1, 1961.

Books

Students and practitioners of O.R. are fortunate in having a hard-core of fundamental works on the subject. All are naturally of recent origin and are readily available. The following list is necessarily selective and does not imply that other literature is less important.

BOWMAN, EDWARD H. and FETTER, ROBERT B. (eds.), *Analyses of Industrial Operations*. Richard D. Irwin, Homewood, Illinois, 1959.

CHURCHMAN, C. WEST, *et al.*, *Introduction to Operations Research*. Wiley, New York, 1957.

EDDISON, R. T., *et al.*, *Operational Research in Management*. English Universities Press, London, 1962.

FLAGLE, CHARLES D., *et al.*, *Operations Research and Systems Engineering*. Johns Hopkins University Press, Baltimore, 1960.

McCLOSKEY, J. F. and TREFETHEN, F. N. (eds.), *Operations Research for Management*. Johns Hopkins University Press, Baltimore. I: 1954, II: 1956.

MORSE, PHILIP M., *Queues, Inventories and Maintenance*. Wiley, New York, 1958.

MORSE, PHILIP M. and KIMBALL, G. E., *Methods of Operations Research*. Wiley, New York, 1951.

RAIFFA, HOWARD and SCHLAIFER, ROBERT, *Applied Statistical Decision Theory*. Harvard Business School, Div. of Research, 1961.
An introduction to the mathematical analysis of decision making.

SAATY, THOMAS L., *Mathematical Methods of Operations Research*. McGraw-Hill, New York, 1959.
A graduate-level work which includes scientific method, mathematical models, optimization, programming, game theory, probability, statistics, and queueing theory. Each chapter has a useful bibliography.

Periodicals

Most of the periodicals which deal specifically with O.R. are published by national societies. They report the results of experiment and practice, review new literature, and give particulars of meetings and other relevant activities. Through their advertisements some of them also provide useful information on job opportunities.

The oldest established scientific journal on the subject is *Operational Research Quarterly*, published for the Operational Research Society (London) by Pergamon Press. The United States counterpart is *Operations Research, the Journal of the Operations Research Society of America*, which is published six times a year

2

and is devoted principally to contributions to the field. The Society also publishes a *Bulletin* twice yearly which includes complete programmes of its national meetings.

Other titles containing a significant amount of O.R. material are:

> *Management Science*, Institute of Management Science, Ann Arbor, Michigan.
> *Naval Research Logistics Quarterly*, Superintendent of Documents, Washington, D.C.
> *Operations Research/Management Science International Literature Digest Service*. Interscience.

The names of additional periodicals may be readily ascertained by perusing the references provided in *International Abstracts in Operations Research*.

Queueing Theory

"Queueing theory is a branch of applied mathematics utilizing concepts from the field of stochastic processes. It has been developed in an attempt to predict fluctuating demands from observational data and to enable an enterprise to provide adequate service for its customers with tolerable waiting." This definition is quoted from Thomas L. Saaty's valuable *Elements of Queueing Theory, with Applications*, which was published by McGraw-Hill, New York, in 1961. The work is arranged in four parts as follows: 1, Structure, technique and basic theory; 2, Poisson queues; 3, Non-Poisson queues; 4, Queueing ramifications, applications, and renewal theory. A most useful feature is an extensive bibliography of 910 items.

Another good bibliography of the subject is Alison Doig's "A bibliography on the theory of queues", *Biometrika*, **44**, 490–514, December 1957. Each of the papers listed is followed by a classification symbol. Capital letters are used to denote the following headings which are reproduced here as they give an insight into some of the applications of the mathematical theory of queues:

C: Problems dealing with storage (content)

F: Problems relating to flow through a network
G: Applications not covered by other categories
I: Inventory problems
M: Problems arising in servicing automatic machines
P: Point processes and counter problems
Q: The general theory of queues
R: Road traffic and related problems
S: Stochastic processes directly related to the study of queues
T: Problems in telephone traffic.

Most of the literature so far published on queueing theory is in non-book form, and it is greatly to the credit of those who have written books on the subject that they have included bibliographical guides. In addition to Saaty there are three other recent works which may be mentioned:

> Cox, D. R. and Smith, Walter L., *Queues*, Methuen, London, 1961. (*Methuen's Monographs on Applied Probability and Statistics*.) Appendix 1 comprises bibliographical notes whose purpose is "to indicate key papers on which the treatment in this monograph is based and from which references to other work may be obtained".
>
> Khintchine, A. Y., *Mathematical Methods in the Theory of Queuing*, Griffin, London, 1960.
>
> Takacs, Lajos, *Introduction to the Theory of Queues*, Oxford University Press, New York, 1962. Each chapter is followed by a bibliography; and the principal mathematical theorems used—Markov chains, Markov processes, recurrent processes—are treated in an appendix.

Theory of Games

The theory of games is another example of a recently developed subject where the bulk of the literature is in the form of papers and periodical articles. Early work was done by John Von Neumann who, together with Oskar Morgenstern, wrote *Theory of Games and Economic Behaviour*, Princeton University Press, 2nd edition, 1947.

A handy introduction is S. Vajda's *An Introduction to Linear Programming and the Theory of Games*, Methuen, London, 1960, in which the theory of games is covered in 26 pages.

Recent works which contain guides to further study are:

> Dresher, Melvin, *Games of Strategy: Theory and Applications*. Prentice-Hall, Englewood Cliffs, New Jersey, 1961.

LUCE, R. DUCAN and RAIFFA, HOWARD, *Games and Decisions*, Wiley, New York, 1957.
A college textbook which includes a very extensive reading list.
McKINSEY, J. C. C., *Introduction to the Theory of Games*, McGraw-Hill, New York, 1952.

Mathematical Programming

Books on mathematical programming differ in treatment according to whether they are aimed at the mathematician or the businessman. On the one hand the theoretical aspects of the subject are emphasized; on the other hand the results which can be achieved by practical application of the theory.

An introduction is provided by Robert W. Metzger's *Elementary Mathematical Programming*, Wiley, New York, 1958. It contains a guide to further reading in which the entries are broadly graded. S. Vajda's *Mathematical Programming*, Addison-Wesley, Reading, Mass., and London, 1961, is a graduate-level textbook of linear and nonlinear programming. Its bibliography contains 128 references.

A most important source book is *Linear Programming and Associated Techniques: A Comprehensive Bibliography on Linear, Nonlinear, and Dynamic Programming* by Vera Riley and Saul I. Gass, Johns Hopkins Press, Baltimore, revised edition, 1958. It comprises references to over 1000 items, including articles, books, monographs, documents, theses and conference proceedings. Part I contains an annotated list of basic references, Part II covers general theory, Part III applications, and Part IV nonlinear and dynamic programming.

"A linear programming problem differs from the general variety in that a *mathematical model* or description of the problem can be stated, using relationships which are called 'straight-line', or linear." This is how Gass puts it in the introduction to his textbook, *Linear Programming: Methods and Applications*, McGraw-Hill, 1958.

The first general linear programming problem was formulated by George B. Dantzig, who also developed the simplex method

K

for its solution. His "Application of the simplex method to a transportation problem" which appears in *Activity Analysis of Production and Allocation* (Tjalling C. Koopmans, ed.), Wiley, New York, 1951, constitutes an important basis of the subject. Subsequent developments are reviewed in Chapter 4 of *Progress in Operations Research*, Wiley, 1961. This survey, by E. Leonard Arnoff and S. Sankar Sengupta, provides references to the relevant literature at every step, and a complete bibliography of the cited works is appended. The succeeding chapter deals with dynamic programming.

The following is a selection of useful titles:

CHARNES, A., *et al.*, *Introduction to Linear Programming*. Wiley, New York, 1953.

FICKEN, F. A., *The Simplex Method of Linear Programming*, Holt, Rinehart and Winston, New York, 1961.

GARVIN, WALTER W., *Introduction to Linear Programming*, McGraw-Hill, New York, 1960.

It is stressed in the preface that this book is an *introduction* to the subjcet and has been limited in size to make it suitable as a textbook for a one-semester course.

GREENWALD, DAKOTA ULRICH, *Linear Programming*, Ronald Press, New York, 1957.

This is intended as an explanatory text for those not possessing advanced mathematics. It is a concise book of only 75 pages,

HADLEY, G., *Linear Programming*, Addison-Wesley, Reading, Mass., and London, 1962.

Assumes a knowledge of linear algebra. The same author's *Linear Algebra* was designed as a complementary volume. *Linear Programming* contains one chapter in which the necessary background material is reviewed.

Exercises

1. From a bibliographical point of view, how can a new science seek to avoid the problems of communication experienced by workers in the older sciences? Illustrate your answer with examples from operational research.
2. What reasons would you give to a person entering the field of O.R. in advising him to join the national society?
3. What steps would you take to ensure that you did not miss the appearance of any important new writings on applications of the theory of games?

Sources of Russian Mathematical Information

Russian-Language Literature

There are two main ways in which collections of Russian literature are built up: by exchange and by purchase. Many libraries and other organizations which can offer publications for exchange enter into agreements with similarly-placed Russian bodies and are thereby able to augment their stocks acquired by purchase.

Periodicals

The purchase of Russian periodicals by subscription is carried out through agents appointed by the central Russian distributing organization, Mezhdunarodnaya Kniga. These agents issue, free of charge, an annual list of available material entitled *Newspapers and Magazines of the U.S.S.R.*

An analysis of Russian mathematical journals appears in *Russian Journals of Mathematics: a Survey and Checklist* by H. A. Steeves, New York Public Library, 1961. It contains three lists as follows: 51 journals which published more than 20 mathematical papers in a specified period; 68 with 7 to 20 papers; and 131 with 3 to 6 papers. There is also a Russian–English glossary.

Books

As for books, the printings are strictly limited, and the supply of any particular title in non-Soviet-bloc countries cannot be guaranteed. Often the entire stock of a book is exhausted in Russia itself, leaving no copies available for foreign booksellers.

Newly published books are listed in the monthly Russian publication *Knizhnaya Letopis'* which has a section devoted to mathematics. Another Russian-language work of value is

Lukomskaya. A. M., *Bibliography of Domestic Literature in Mathematics and Physics*, Akad. Nauk. S.S.S.R., Biblioteka, Moscow–Leningrad, 1961. It contains an appendix which lists sources of bibliographical information.

Two volumes published in the United States which may be examined are:

> FORSYTHE, GEORGE ELMER, *Bibliography of Russian Mathematical Books*, Chelsea Publishing Co., New York, 1956.

and

> LA SALLE, JOSEPH PIERRE, and LEFSCHETZ, SOLOMON, *Recent Soviet Contributions to Mathematics*, Macmillan, New York, 1962.

Guides to Collections

The best English-language publication for keeping abreast of Russian mathematical literature is undoubtedly *Mathematical Reviews*, which is described in Chapter 4. However, in order to ascertain the availability of any particular document, recourse has to be made to union catalogues, holdings lists, and library accessions lists.

The *Monthly Index of Russian Accessions*, published by the Library of Congress, is a record of publications in the Russian language received by the Library of Congress and a group of co-operating libraries. It is arranged in three parts:

> Part A: Monographic Works. (In each entry, the title in the original language is preceded by its English translation in brackets.)
> Part B: Periodicals. (Simply indicates which issues of each title have been received. The entries are arranged in subject groups.)
> Part C: Subject Index to Monographs and Periodicals. (This is the largest section and is an extremely valuable analytical index. Items are arranged under subject headings, and the whole is preceded by a list of periodicals indexed and abbreviations used.)

Various cumulations are available, and orders for microfilm or photostat copies of the listed items which are in the collections of the Library of Congress may be placed with the Library's Photo-duplication Service.

Another valuable Library of Congress guide is its *Serial*

Publications of the Soviet Union, 1937–1957: a Bibliographic Checklist.

In Britain, the National Lending Library of Science and Technology (N.L.L.) is rapidly expanding its collection of Russian literature. Book accessions are announced in its *Monthly List of Books Received from the U.S.S.R. and Translated Books*, which is available free of charge on request. The list is arranged in broad subject groups, approximately corresponding to the Universal Decimal Classification. The N.L.L.'s translation activities are described in the following section.

Literature in Translation

General Guides

Since only a relatively small number of mathematicians in Western countries are able to read Russian, it has become imperative for translation on a large scale to be undertaken if important Russian contributions are to be fully exploited. Until a few years ago Western scientists remained largely unaware of the scientific progress being made in the Soviet Union. Today, however, there is little excuse for not keeping informed. Translation activity is so widespread and virile that it has become necessary to produce guides to its ramifications.

One of the first of these bibliographical signposts was entitled *Providing U.S. Scientists with Soviet Scientific Information.* It proved useful enough to warrant a second edition in 1962. Compiled by B. I. Gorokohoff and published by the Massachusetts Institute of Technology, it covers such things as translations of periodicals, books and other documents, abstracting services, and the availability of this material in United States libraries.

An international survey is "Translations of Russian scientific and technical literature" by Alice Frank, *Revue de la Documentation*, **28**, 47–51, May, 1961. The areas dealt with are Austria, Belgium, Canada, France, Germany, Netherlands, Scandinavia, Spain, United Kingdom and the United States.

Also worth noting is *Notes on Searching Russian Scientific*

Literature in Translation by Harry La Plante, published by the University of Detroit Library in 1961.

Government Support

In the United States, the Office of Scientific Information Services of the National Science Foundation is particularly concerned in this field. It was established in 1958. Two of its designated programmes are Support of Scientific Publications and Foreign Science Information. The Office supports, through grants, the cover-to-cover translation of some 50 periodicals and a large number of Russian books. Recently, for example, it awarded $16,687 for the translation of *Mathematics: its Contents, Methods and Meaning* by A. D. Aleksandrov *et al*. With N.S.F. support the Battelle Memorial Institute is preparing a guide to East European scientific and technical literature available to U.S. scientists, and a directory of scientific institutions in the U.S.S.R., which will include details of personnel and publications.

British government activity is centred on the Department of Scientific and Industrial Research, which operates the National Lending Library of Science and Technology. Anglo-American co-operation is firmly established. The National Lending Library Loan Collection of Translations includes microfilms of all Russian translations held by the Special Libraries Association at the John Crerar Library in Chicago. These are listed in the *Bibliography of Translations from Russian Scientific and Technical Literature* (Library of Congress, 1953–6), and *Translations Monthly* (John Crerar Library, Volumes 1–4, 1955–8). Current accessions are listed in *Technical Translations* issued by the Office of Technical Services of the U.S. Department of Commerce. The National Lending Library issues, each month, its own *N.L.L. Translations Bulletin* (H.M.S.O., London) which reports on a variety of translation activities.

Translation Services

Translators and Translations: Services and Sources, first published by the Special Libraries Association, New York, in 1959 and since revised, offers a unique guide. A register is also maintained by the American Translators Association. In Britain, such bodies as the Institute of Linguists and Aslib can advise on mathematical translation services.

Translation of Periodical Literature

Over 2500 Russian scientific and technical periodicals are published currently. The information they contain is being made increasingly available to English-speaking scientists and technologists through translation and review services of many kinds. These range from complete cover-to-cover translations of important periodicals to individually commissioned translations of specific items.

Translations are initially expensive, but through co-operative enterprises costs have been shared, thus making the translated material more readily accessible.

Well over a hundred Russian periodicals are now being completely translated into English on a continuing basis. Many of the publishers receive government subsidies to enable them to maintain subscription rates at a reasonable level.

"Cover-to-cover translations", according to Aslib, "have two main functions. On the one hand they help to overcome ignorance of foreign scientific literature generally, to spread awareness of its importance, and stimulate a demand for it. For languages which few people can read this function is particularly important. At the same time they provide in published form translations of individual articles which some people wish to read and should thus reduce the number of translations which might otherwise be made specially by or for the individuals who want them." (*The Foreign Language Barrier*, Aslib, London, 1962.)

Lists of such translations are published in *N.L.L. Translations Bulletin* (see p. 132) and in the Library of Congress *Monthly Index of Russian Accessions* (see p. 130).

The following is a selection of titles and suppliers of those journals which are of particular interest to mathematicians:

Automation and Remote Control
  (*Avtomatika i Telemekhanika*)
  Instrument Society of America,
  313 Sixth Avenue,
  Pittsburgh 22, Pennsylvania, U.S.A.
Journal of Applied Mathematics and Mechanics
  (*Prikladnaya Matematika i Mekhanika*)
  Pergamon Press, Ltd.,   Pergamon Press, Inc.,
  4 Fitzroy Square,   122 East 55th Street,
  London, W.1, England   New York 22, New York, U.S.A.
Problems of Cybernetics
  (*Problemy Kibernetiki*)
  Pergamon Press, Ltd.
  (as above)
Progress in the Mathematical Sciences; Selected Articles
  (*Uspekhi Matematicheskikh Nauk*)
  Cleaver-Hume Press, Ltd.,
  31 Wright's Lane,
  London, W.8, England
Soviet Mathematics—Doklady
  (*Akademiya Nauk S.S.S.R., Doklady - Otdel. Matematiki*
  Proceedings of the Academy of Sciences of the U.S.S.R.,
  Mathematics Section)
  American Mathematical Society,
  190 Hope Street,
  Providence 6, Rhode Island, U.S.A.
Soviet Physics—JETP
  (*Zhurnal Eksperimental'noi i Teoreticheskoi Fiziki*
  Journal of Experimental and Theoretical Physics)
  American Institute of Physics,
  335 East 45th Street,
  New York 17, New York, U.S.A.
Theory of Probability and its Applications
  (*Teoriya Veroyatnosti i ee Primeneniye*)
  Society for Industrial and Applied Mathematics,
  Box 7541, Philadelphia 1, Pennsylvania, U.S.A.
U.S.S.R. Computational Mathematics and Mathematical Physics Journal
  (Selections from *Zhurnal Vychislitel'noi Matematiki i Matematicheskoi Fiziki*)
  Pergamon Press Ltd.
  (as above)

Translation of Monographic Literature

There is a plentiful supply of good mathematical books published in the U.S.S.R., though it is not always easy to obtain copies outside the Soviet bloc countries. For the most part they are very cheap. Every effort is being made to produce translations of those which are considered to have the greatest value. Two sources of information on such translations were noted in Chapter 3; they are the "Mathematics in translation" section of the American Mathematical Society's *Notices* and UNESCO's *Index Translationum*. A third is *Russian Technical Literature* published by the Organisation for Economic Co-operation and Development in Paris. This covers Russian and other Eastern scientific and technical publications. It is concerned with sources of information, translation centres, dictionaries, translation techniques, and actual translations completed or in progress.

Book translations are made either as part of a planned publishing programme or as individually commissioned work. Two examples of the former are the Israel Program for Scientific Translations, which includes mathematics, and the scheme of the Hindustan Publishing Corporation of Delhi, India, under which a number of English translations of Russian mathematical books have been marketed. These are by no means the only producers of translations, many reputable scientific publishers such as Pergamon Press and Consultants Bureau being active in the field.

Russian–English Mathematics Dictionaries

More English-speaking scientists than ever are learning Russian as the educational facilities increase and suitable textbooks are produced. The mathematician is at somewhat of an advantage. The vocabulary of each branch of his subject is to some extent international. It is relatively limited and highly specialized. With a good knowledge of the subject, a reading ability of mathematical texts in the important languages is not as difficult as might at first be imagined.

In dealing with Russian, the Cyrillic alphabet has first to be mastered, but a little practice in transliteration (converting Russian characters into English characters) soon renders this less frightening. There are several different systems of transliteration which vary slightly in certain respects.

A knowledge of the order of the letters in the Cyrillic alphabet is necessary for locating Russian works in a dictionary.

Many Russian–English dictionaries have been produced in the past few years to meet the rising demand. A New York University study which includes the dictionary needs in the fields of mathematics and allied subjects is *Russian–English Scientific and Technical Dictionaries—a Survey*. This is available from Professor A. F. Hubbell, Gallatin House, New York University, New York 3, New York, U.S.A.

A *Russian–English Dictionary of the Mathematical Sciences*, compiled and edited by A. J. Lohwater with the collaboration of S. H. Gould, was published by the American Mathematical Society in 1961. It includes a concise grammar of the Russian language.

Also available are the *Mathematical Dictionary: Russian and English*, published by the Deutscher Verlag der Wissenschaften, Berlin, in 1959 (see p. 29), and Louis Melville Milne-Thomson's *Russian–English Mathematical Dictionary*, published by the University of Wisconsin Press in 1962.

Abbreviations are commonly used in Russian works and can often cause difficulty. This applies both to the actual subject matter and to the names of institutions. The Scientific Research Institute of Mathematics, for instance, is quoted as N.I.I.M. This stands for Nauchno-Issledovatel'skii Institut Matematiki. Such contingencies are catered for by *Russian Abbreviations*, *a Selective List* compiled by Alexander Rosenberg, Library of Congress, Washington, D.C., 2nd edition, 1957.

Mathematics and the Government

Introduction

The cost of research in science and technology mounts steadily every year. It is borne by every sector of the community, either directly as taxation or indirectly in the price of goods—the end products of research. The three principal sponsors of research are industry, universities and governments, though this is an extreme simplification of the picture. Often, the constituent members of a particular industry co-operate to form research associations through which the cost of research can be shared. Co-operation can be international as well as national, and the individual participating organizations vary considerably in size. Generally speaking, the idea of free exchange of information is becoming much more widely accepted as beneficial to all parties concerned, though of course the practice of patenting new inventions continues to flourish as vigorously as ever.

A unique example of altruism is to be seen in the Battelle Memorial Institute in the United States. The Institute, which carries out a very substantial volume of research, was expressly founded by the will of Gordon Battelle as a practical means for producing social benefits through scientific research and the making of scientific discoveries and inventions. Every month it issues the *Battelle Technical Review* which reports its activities and contains abstracts of an international range of technical documents reflecting the primary areas of its research. Among these are control systems, computers, automation, cybernetics, systems engineering, and operational research.

In Great Britain there are some 50 research associations which

carry out research projects for member organizations, but whic]
also make available to the public at large the results of the greate
part of their research activities. Their organization is describe
below. For the present, it will be noted that they have stron
financial backing from the British Government. Governmen
support does in fact pervade the whole domain of scientif
research. It can be seen in the financial backing of researc]
publications, in the placement of research contracts in industry
in the maintenance of its own research laboratories, in grants t
universities, and in many other forms. Some of these activities
and in particular those which affect mathematics, are worth
closer look.

United States

The activities of the majority of U.S. Government department
and agencies are described in the *United States Governmen
Organization Manual*, which is published annually by the Genera
Services Administration. This is a most useful source for obtain
ing the addresses of offices whose names only are usually quotec
in articles and references. A guide to U.S. Government publica
tions themselves is *Government Publications and Their Use*, b
Laurence F. Schmeckebier and Roy B. Eastin. It was first pub
lished in 1936, but a revised edition was published by The Brook
ings Institution, Washington, D.C., in 1961.

The government maintains several research groups of its own
which, by designation, are mathematically orientated. The
include the following:

> Air Force Applied Mathematics Research Branch at the Wright Aero
> nautical Development Center
> Office of Naval Research Mathematical Sciences Division
> Naval Ordnance Laboratory Mathematics Department
> David Taylor Model Basin Applied Mathematics Laboratory, U.S. Navy
> Army Research Office Mathematical Sciences Division.

In 1961, the U.S. Naval Research Laboratory also announced
the creation of an Applied Mathematics staff as part of the Office

of the Director of Research, primarily to carry out research programmes on numerical analysis, mathematical physics, and
optimization techniques. Recently, the Air Force Office of
Scientific Research expanded its support for research in applied
mathematics. Details are available from the Directorate of
Mathematical Sciences, Air Force Office of Scientific Research,
Washington 25, D.C.

In addition, numerous federal research contracts are placed
with industrial companies, universities and other organizations
equipped for research or the publication of research results.

U.S. Government Research Reports is issued twice a month by
the Office of Technical Services of the U.S. Department of
Commerce. It announces new reports of research and develop-
ment released by the Army, Navy, Air Force, Atomic Energy
Commission, and other agencies of the Federal Government.
Information on where the original documents may be obtained is
given. Normally they are available prepaid direct from the
Office of Technical Services. Microfilm copies may also be obtained
on request. The publication is in two sections. The first is called
the "Technical Abstract Bulletin" (compiled by the Armed
Services Technical Information Agency), and the second "Non-
Military and Older Military Research Reports". Both sections
include mathematics headings.

U.S. Government Research Reports does not, however, provide
an exhaustive coverage of reports prepared as a result of government contracts. Another useful source of current information is
the *Monthly Catalog of U.S. Government Publications*.

Twelve new Regional Technical Report Centers have been set
up in universities and libraries with the responsibility of making
available the results of government-sponsored research and
development. Each centre receives the reports of the Department
of Defense, the National Aeronautics and Space Administration,
and the Atomic Energy Commission, and meets demands from
the public through reference, lending, and photocopying facilities.
The centres are:

Georgia Institute of Technology (Atlanta)
Massachusetts Institute of Technology (Cambridge)
John Crerar Library (Chicago)
Southern Methodist University (Dallas)
University of Colorado (Boulder)
Linda Hall Library (Kansas City)
University of California (Los Angeles)
Columbia University (New York City)
Carnegie Library of Pittsburgh
University of California—Berkeley (San Francisco)
University of Washington (Seattle)
Library of Congress (Washington, D.C.).

The activities of the National Bureau of Standards are of particular interest to the mathematician. They are divided among various divisions and sections including Applied Mathematics, which deals with numerical analysis, computation, statistical engineering, mathematical physics, and operational research. Its *Journal of Research* is published quarterly in two parts. Section B is entitled *Mathematics and Mathematical Physics*. It reports studies and compilations designed mainly for the mathematician and the theoretical physicist, and each issue contains abstracts of the Bureau's publications. Particularly valuable is a series of bibliographies dealing with specific topics in the field of statistics, under the general title: "Selected bibliography of statistical literature 1930–1957". The entries are drawn from a file of abstracts currently maintained in the N.B.S. Statistical Engineering Laboratory (see p. 109).

The National Science Foundation is noted elsewhere in this book with reference to the financial support it gives to mathematical publications, especially translations of foreign-language works. It was established by Act of Congress in 1950, and its work is described in a pamphlet, *Program Activities of the National Science Foundation* (N.S.F., 1951 Constitution Ave., Washington D.C.). The two divisions of most interest to mathematicians are the Division of Mathematical, Physical, and Engineering Sciences and the Division of Scientific Personnel and Education. Information on N.S.F. grants and contracts in support of scientific information is given as a regular feature of its publication, *Scientific Information Notes*.

A national library service is provided by the Library of Congress whose printed author and subject catalogues are highly valued bibliographical tools. The Science and Technology Division issues specialized publications, one of which is a *List of Russian Serials Translated into English and Other Western Languages*.

The U.S. Department of Health, Education, and Welfare is another relevant government department. An example of its publications is *Facilities and Equipment for Science and Mathematics: Requirements and Recommendations of State Departments of Education*, 1960.

Details of fellowships and other support for basic research in mathematics are listed in *A Selected List of Major Fellowship Opportunities and Publications for Educational Research*, which is available from the Fellowship Office of the National Academy of Sciences—National Research Council (2101 Constitution Avenue, Washington 25, D.C.).

A book worth noting is Kidd, Charles V. *American Universities and Federal Research*, Belknap Press of Harvard University Press, 1959.

United Kingdom

The Department of Scientific and Industrial Research operates several research *establishments* of its own, and in addition supports the work of some 50 research *associations* serving the needs of particular groups of industries. Their activities are described respectively in the annual reports of the research establishments and in *Research for Industry*, which covers the research associations. In 1962, Her Majesty's Stationery Office published *D.S.I.R., Universities and Colleges 1956–60: a Report on D.S.I.R. Support for Research and Training in Universities and Colleges*. Among other things, this report gives details of grants made to individual students. D.S.I.R. also issues current data on work in progress in *Research in British Universities*. The entries are arranged in alphabetical order of institutions which include colleges of technology, universities, and university colleges. The arrangement within each entry is an alphabetical list of subjects. Specific areas of research are given under these subject headings, together with the names of departmental heads and the staff members engaged in supervising research. There is a complete name index and a detailed subject index.

Industrial Research in Britain, Harrap, London, 4th edition, 1962, is a very comprehensive work dealing with all aspects of the subject from sponsored research organizations to periodical abstract journals covering industrial research. There is also an article on "The Government and industrial research" by Sir Harry Melville, Secretary of the Department of Scientific and Industrial Research.

The National Lending Library for Science and Technology, in Boston Spa, Lincolnshire, became fully operational in 1962. It meets requests for loans and photocopies from a vast stock of literature which includes the largest collection of Russian serials in Western Europe. The extensive loan service previously operated by the Science Museum Library in London has been absorbed into the National Lending Library's activities. A counterpart library offering reference services is planned.

British government publications are made known in the *Daily List*, the *Monthly Catalogue*, and the *Annual Catalogue*, all issued by H.M.S.O. A five-yearly index is also published. *Sectional Lists*, which are available free on request, are of a more specialized character. Typical of them is that for the Department of Scientific and Industrial Research (*Sectional List* No. 3) of which a useful feature is that most of the publications it includes are graded as follows: "(A)—Suitable for an intelligent reader who has no previous knowledge of the subject, (B)—Suitable for a senior student or practitioner of the subject, and (C)—Suitable for research workers and specialists in the subject". A list of D.S.I.R.'s research and other establishments is also given, together with details of their publications. An important entry under the

National Physical Laboratory heading, for example, concerns its *Mathematical Tables Series*, the first of which is entitled *The Use and Construction of Mathematical Tables*. When the series of *Sectional Lists* is complete it will comprise a catalogue of all current parliamentary publications. The lists are periodically revised.

Title, frequency and price of government periodicals are given in the printed catalogues.

In the United States, British government publications are obtainable from British Information Services, 45 Rockefeller Plaza, New York 20, New York. A complete list of overseas agents appears in the H.M.S.O. catalogues mentioned above. In Britain there are government bookshops in seven major cities, and their names and addresses together with booksellers who act as agents for government publications are similarly given in the catalogues.

Government Information and the Research Worker, edited by Ronald Staveley, and published by the Library Association, London, in 1952, contains a chapter dealing with H.M.S.O. and its publishing policy. Other chapters are contributed by experts on each of the major government departments. A new edition is in preparation.

L

Actuarial Science

Orientation

In the Dewey Decimal Classification, actuarial science is provided for in the Insurance class 368. The ramifications of the subject are, however, very wide. Its roots go deep into the theories of probability and statistics. Mathematics is, indeed, an essential part of the professional actuary's background. A brief summary of sources of relevant information will therefore be given.

The following definition is taken from a useful booklet entitled *The Actuarial Profession*, published by the Institute of Actuaries in London, 1962:

> *Actuarial science is concerned with applying the theory of probability, and statistical processes generally, to practical affairs and especially to the financial problems connected with the management and administration of life insurance, pension schemes and social insurance.*

It follows that since much of the actuary's work involves statistics, the bibliographical sources described in Chapters 11 and 12 will be of considerable value.

Again, attention should be drawn to Dewey's Statistics class 310, the broad arrangement of which is as follows:

310: Statistics
  e.g. United Nations, *Statistical Yearbook*, New York.
311: Statistical method
  e.g. Snedecor, George W., *Statistical Methods*, Iowa State College Press, Ames, Iowa, 5th edition, 1956.
312: Demography
  e.g. Cox, Peter R., *Demography*, Cambridge University Press, 3rd edition, 1959.

314–319: General statistics of specific countries
e.g. 314.2 Central Statistical Office, *Annual Abstract of Statistics*, H.M.S.O., London.
317.3 U.S. Bureau of the Census, *Historical Statistics of the United States, Colonial Times to* 1957, Washington, D.C., 1960.

Monographic Literature

In the absence of relevant bibliographies the monographic literature of actuarial science is best represented in the catalogues of libraries devoted to the subject. They inevitably reveal a pre-ponderance of statistical works. This is clearly shown in the library of the Institute of Actuaries, London, which has a stock of some 10,000 volumes. Its *Additions to the Library* is issued as a separate booklet and distributed with the *Institute of Actuaries Year Book*. The latter itself includes a reading list for students. It is graded insofar as the references are related to the various examinations held by the Institute. The Institute publishes its own series of student textbooks which are available to members at preferential rates. Recent volumes, quoted as examples only, include *Finite Differences for Actuarial Students* by H. Freeman, and *Probability, an Intermediate Text-Book* by M. T. L. Bizley. Another valuable work, *Actuarial Statistics*, is in two volumes: 1, *Statistics and Graduation*, by H. Tetley (1950); and 2, *Construction of Mortality and Other Tables*, by J. L. Anderson and J. B. Dow (1958).

Government statistical publications are of great importance, and the guides described in Appendix II will be found useful in locating them. Two series which may be noted are the British *Government Actuary's Reports*, and the *Actuarial Studies* of the U.S. Department of Health, Education and Welfare.

Foreign actuarial terms are included in the *International Insurance Dictionary* published by the European Conference of Insurance Supervisory Services in Berne, 1959. As indicated in the preface, the international actuarial notation is a reliable link between different languages. Part IV of the Dictionary is therefore devoted to actuarial symbols with explanations of their meaning in English, German, Dutch, French, Italian, Spanish, Portuguese, Danish, Swedish, Norwegian, and Finnish.

Periodicals and Abstracts

Due to the diversity of subjects in which the actuary must achieve competence, his breadth of reading is quite substantial. In addition to purely actuarial journals he has considerable interest in those relating to mathematics, statistics, and computation. More recently his list has been increased to cover such fields as operational research. The following English-language titles would therefore constitute a fairly representative list:

> *American Mathematical Monthly; Annals of Mathematical Statistics; Annals of Mathematics; Applied Statistics; Computer Journal; Journal of the American Statistical Association; Mathematics of Computation; Operational Research Quarterly;* and *Operations Research.*

His basic English-language actuarial publications include:

> *Journal of the Institute of Actuaries,*
> *Transactions of the Actuarial Society of Australasia,*
> *Transactions of the Faculty of Actuaries,* and
> *Transactions of the Society of Actuaries.*

World progress in the subject, and reviews of foreign-language publications, are well covered by the above reading. For more detailed information on activities in other countries, however, the actuary may have recourse to the periodical publications of the respective national societies.

There is no central abstracting journal covering actuarial science, but by regularly consulting two or three more general services—especially in the field of statistics—the actuary can keep informed quite adequately. Some actuarial journals go much further than simply providing book reviews. The *Journal of the Institute of Actuaries* (Alden Press, Oxford), for example, includes three regular sections to assist its readers in keeping abreast:

1. Notes on other actuarial journals (covering the overseas press).
2. Notes on the *Transactions of the Faculty of Actuaries.*
3. Articles, papers and publications of actuarial interest (comprising annotated references to appropriate items selected from a wide range of peripheral publications).

Actuarial Societies

International

The profession's international organization is called the Comité Permanent des Congrès Internationaux d'Actuaires (9 rue des Chevaliers, Brussels, Belgium). The first International Congress of Actuaries was held in Brussels as far back as 1895. Congress papers are published in a volume of *Proceedings*.

ASTIN is a section of the Comité which was established in 1957 in New York. Its purpose is the promotion of mathematical research in non-life insurance, and its periodical publication is called *Astin Bulletin*.

National

National bodies contribute to the work of the Comité as well as acting as co-ordinating centres for their respective countries.

The Institute of Actuaries (Staple Inn Hall, High Holborn, London, W.C.1), working in close co-operation with the Faculty of Actuaries in Scotland, is not only the focal point of the profession in England, but also in the Commonwealth. It was founded in 1848, and incorporated by Royal Charter in 1884. Its publications have already been described, and its educational and career activities are dealt with in Chapter I.

In the United States, the national body is the Society of Actuaries (208 South La Salle Street, Chicago 4, Illinois). Founded in 1949, it absorbed both the Actuarial Society of America and the American Institute of Actuaries. It has active meetings and publishing programmes, and sponsors Associateship and Fellowship examinations.

The following is a list of other national actuarial societies, arranged alphabetically by country:

Australia:	Actuarial Society of Australia and New Zealand,
	c/o Australian Mutual Provident Society,
	87 Pitt Street, Sydney, N.S.W.
Belgium:	Association Royale des Actuaires Belges,
	74 rue Royale, Brussels.
Canada:	Canadian Institute of Actuaries
	302 Bay Street, Toronto.

France:	Institut des Actuaires Français, 247 rue Saint Honoré, Paris.
Germany:	Deutsche Gesellschaft für Versicherungsmathematik, Von Werth Strasse 4–14, Cologne.
India:	Actuarial Society of India, c/o Life Insurance Corporation of India, Oriental Building, Mahatma Gandhi Road, Fort, Bombay 1.
Italy:	Istituto degli Attuari, Via dell' Arancio 66, Rome.
Scotland:	Faculty of Actuaries in Scotland 23 St Andrew Square, Edinburgh.
South Africa:	Actuarial Society of South Africa, P.O. Box 4464, Cape Town.
Switzerland:	Association des Actuaires Suisses, Aeschemplatz 7, Basel.

Index

149